Healing the Biosphere: A Scientific Imperative

Alan P. Ammann, PhD

Healing the Biosphere Copyright © 2018 by Alan P. Ammann. All Rights Reserved.

All rights reserved. No part of this book may be reproduced in any form or by any electronic or mechanical means including information storage and retrieval systems, without permission in writing from the author. The only exception is by a reviewer, who may quote short excerpts in a review.

Printed in the United States of America

First Printing November 2018

ISBN- 9781730817748

Contents

INTRODUCTION	1
WHAT IS A HEALTHY BIOSPHERE?	7
THE LIE WE TELL OURSELVES	9
AN OUTSIDE OPINION	12
THE TASKS WE FACE	15
SCIENCE, RELIGION AND POLITICS	18
RESPONSIBILITIES OF ENVIRONMENTAL SCIENTISTS	29
RESPONSIBILITIES OF RELIGIOUS LEADERS	33
RESPONSIBILITIES OF POLITICAL LEADERS	36
FACING REALITY	39
THE FACTS OF LIFE	41
THE EVOLUTION GAME	46
ECOSYSTEMS, NATIVE AND OTHERWISE	55
THE BIOSPHERE IN REALITY	66
THE BIOSPHERE IN OUR MINDS	69
WHY WE ARE DESTROYING THE BIOSPHERE	78
OUR BUBBLE OF DENIAL	84
VALUING THE BIOSPHERE	87
A RELIGIOUS MORAL IMPERATIVE	101
A SECULAR MORAL IMPERATIVE	119
SCIENCE AS A TOOL FOR HEALING THE BIOSPHERE	122
EXPECT A RELIGIOUS BACKLASH	149
APPENDIX 1: MYTHS THAT HARM THE HEALTH OF THE BIOSPHERE	152

Questioning our Sacred Myths 152
Sacred Texts Contain Revealed Truth about the World 153
God or Other Divine Entities Created the World. 154
The Rules of Evolution and Ecology don't apply to Humans 156
Engineering can Replace Ecosystems 158
We have Dominion over the World 159
Be Fruitful and Multiply 160
The World is a Temporary Stop on the Way to Eternity 161
The end of the World is at Hand 163
The Personification of the Biosphere (E.g. Mother Nature) 164
The World is Illusion 166
The Magical Thinking of Prayers and Incantations 167
Living in Harmony with Nature 167
My God is the only God 168
Humans are innately good: Humans are innately bad 169
The wisdom of Free Markets 169

ABOUT THE AUTHOR 171

INTRODUCTION

The Biosphere is our global ecosystem, that thin layer of the earth's crust and atmosphere in which all known life exists and on which it depends. It is the arena in which life arose and continues to evolve. While it is very robust, having supported life for more than 3 billion years through many catastrophic events, the cumulative impacts of human activities over the last several millennia are rapidly destroying its health. Our foolishness is leading to a world more suited to rodents, roaches and other species we consider vermin, than ourselves.

If we degrade the Biosphere to the point where it no longer supports us, we have nowhere else to go. A few of us might live on in small engineered shelters, but civilization as we know it would end. Any survivors struggling to restore civilization would be doing so in a devastated world. The difficulties of rebuilding civilization would be compounded by the loss of scientific knowledge and its likely replacement by the same sort of myths and superstitions that promoted its devastation in the first place.

The truth is we cannot terra-form Mars and humanity cannot escape to another solar system, that only happens in science fiction. For better or worse Earth's Biosphere, the one we take for granted and treat so carelessly, is our only home, the only place in the known universe where humanity can live.

I am not an alarmist although what I have to say about the health of the Biosphere is alarming. I have studied and worked as a Wildlife Biologist for over 50 years, having spent nearly thirty of those years with a Federal Agency. During much of that career I worked in the field on ecosystem restoration, primarily salt marshes and other habitats in New England.

From my personal experiences I can say that being an Ecosystem Restorationist shares some aspects of being the forward watch on the Titanic. You see the iceberg and shout a warning but in the main salon the party goes on. Even after the collision arguments ensue about how bad it is. Finally, reality sets in and it turns out there are not enough life boats. The point being that it is not enough to know of imminent danger if you can't convey the warning to others, and the warning if heeded, must be heeded in time or the ship will sink. Unfortunately, like the passengers and crew of the

Titanic, humanity is better at arranging deck chairs than avoiding ecological icebergs.

This book is my shouted warning from the forward lookout post. But not just another warning, rather its purpose is to examine why we are degrading the Biosphere at such an alarming rate and to suggest how we might change course before we hit an environmental iceberg. It is an extended essay, some might say a polemic, based on my education and experience. While not a scientific document *per se*, I believe it is based on sound science. All I ask is that those who read this book consider what I say with that facility, unique to humans, reason. For it is through the cool light of reason not the hot light of passion I hope to illuminate a path to the future in which we as a species can live a life of purpose and dignity in the company of our fellow creatures on a healthy planet.

My central thesis is that we are not depleting the Biosphere because we are an inherently evil species, rather we are simply doing what we are programmed to do by the process of biological evolution. In this we are the same as all other organisms on the Earth. We use our intelligence and manual dexterity instinctively as the tiger uses its fangs and claws. The only difference is that the power of our evolutionary adaptations coupled with our gross overpopulation allows us to exploit the

Biosphere orders of magnitude beyond its carrying capacity. We are a keystone species run amok; a species with intelligence, not an intelligent species.

Changing human behavior from thoughtless exploitation of the Biosphere to accepting our responsibility to heal it will not be an easy task, if it can be done at all. But one thing is clear, accepting the status quo is not an option if human civilization is to have a future. We must act on a global scale while we still have the time, but we must act in a reasoned and deliberate way. It will require the efforts of our best minds and international cooperation on an unprecedented scale.

Current conservation efforts are not adequate, either in commitment or scale. The stark truth is that for all the talk about being green very little has been done to heal the Biosphere on a global scale. It will take much more than creating few wildlife reserves and recycling a few bottles and cans to save us. Drastic changes will be necessary that will affect the very foundation of human society from government too religion.

Simply put: A healthy Biosphere is a human right and humanities responsibility. Enough of humanity to matter must come to believe that we have a mandate from our creator, whether that be a divine entity or a collection of indifferent natural processes, to heal and maintain the Biosphere. Restoring the health of the Biosphere would be one of the largest and most expenses projects in human history. What I suggest is that healing the Biosphere will not take place based only on the fact that it is a scientific necessity. It must become a moral imperative as well.

I recognize that there are numerous barriers to an undertaking of this magnitude. Too many are ignorant or in willful denial that such a problem exists. What I propose is a strategy to restore the health of the Biosphere through the integrated use of Science and Sacred Myth. Science would tell us what to do and myth would provide the moral imperative to do it. Whether or not we have the necessary will to make the required sacrifices I do not know. I, for one, cannot and will not accept that our self-extinction is either divinely ordered or an evolutionary certainty.

Despite all that I know, I am an incorrigible optimist. I believe that there will come a time, if we can survive our present situation, when most will accept that we are the children of the Biosphere, it is our Mother, Father and Creator. Our scientific understanding of it is as close to divine revelation as we are likely to get. Even though it is an impersonal natural process governed by the rules of Evolution, we owe it our reverence and love because that is how the world is and this life as a human being on planet earth is the only life we get. And if we have a purpose in this universe it is to prove that intelligence is a successful evolutionary adaptation.

I wish that I was not compelled by conscience to write this book. I would much rather be tending my garden in peaceful retirement as Voltaire suggests in "Candide". If what I write contributes to the healing of the Biosphere, I will consider this book a success. I take full responsibility for its contents and apologize for any errors and omissions.

Alan P. Ammann, Ph.D.

WHAT IS A HEALTHY BIOSPHERE?

Because this book is about healing the Biosphere I will begin with my definition of a healthy Biosphere. A healthy Biosphere would be a complex, interconnected, biologically diverse, global ecosystem, in which our species can live and evolve is such a manner that our activities do not stress it beyond its carrying capacity to provide the basic needs, e.g. oxygen and a livable climate for ourselves and our fellow organisms. A healthy Biosphere, on this or any other planet, could take many forms depending on such things as climate, geology, soils, atmospheric composition, the number and types of extant species and so forth.

A healthy Biosphere would continually change over time due to the forces of biological evolution and subject to cosmic and geological processes and events, such as meteor strikes, volcanos and continental drift. What is important is that we restore the Biosphere as a fully functioning sustainable global ecosystem, one that has enough biodiversity to sustain us in the long term. How long can such a Biosphere last? I have no idea, but I can say that it will last longer than the Biosphere in its present condition.

I include our presence in the definition of a healthy Biosphere because life can be understood as both an objective process and a subjective experience; life as a biologist sees it, a complex process of biological evolution played out in the Biosphere and our life as we live it, the day to day experience of a sentient and sapient organism on planet Earth.

Objectively, humans are not a necessary ingredient of a healthy Biosphere. The Biosphere would have chugged merrily along if we had never evolved. But we did evolve and subjectively, like all species, we instinctively strive to exist, it is the natural order of things. From our perspective, the Biosphere must be healthy subjectively as well as objectively. It must satisfy our emotional as well as our physical needs. Nature must be beautiful as well as functional. We may not be essential to the Biosphere when viewed as an objective process, but when viewed from our perspective as a subjective experience we are free to interpret our presence as the premier sentient and sapient creature in it to be the meaning and purpose of it all.

THE LIE WE TELL OURSELVES

Plato is reputed to have said "I thank God that I was born Greek and not barbarian, free and not slave, male and not female, but above all that I was born in the age of Socrates." What he did not say was "thank God that I was born human and not an animal.". He did not need to say this, it was assumed that we are divinely created different from all other animals. He was born into this lie as we are, no matter what race or culture.

I believe that humanity will not begin the task of healing the Biosphere without the leadership of those who understand that the very concept of humanity is a lie we tell ourselves. I do not mean that Plato was a liar, simply that he was caught up in a cultural lie. It must have been obvious, even to the ancients that our lives are like other animals, we eat, sleep and reproduce and are made of flesh and blood. Indeed, they saw people's insides following a battle or accident and even to the untrained eye, our viscera, look very much like an ox or rabbit. But seemingly, from the beginning of sentience we came to believe that we had something that other creatures did not, something that made us different. The problem was that no one could see what that something was. Lacking any visible evidence, the ancients invented a difference, an invisible soul implanted in us by an unseen deity, a ghost in the machine that is the real us and may live on after the death of our physical body.

The need to believe this lie seems to be instinctive, it has been passed down in all cultures and religions and expressed in every known language. We want to believe that animals are animals and humans are humans, despite all evidence to the contrary. Science shows us that the true repository of our being is a set of chromosomes not a soul and that we are subject to the same laws of evolution as every other species. Admittedly, we have a higher level of sentience and some would argue sapience than other animals. But these attributes are themselves evolutionary adaptations we share with many other species. What does make us unique is the prosaic fact that our evolutionary adaptations, our intelligence and manipulative skills, are orders of magnitude more powerful than those of other species. Our differences from other animals are qualitative not quantitative; we differ in degree not in kind.

I will use the following framework for understanding why we are in the processes of seriously degrading the Biosphere and explaining my thesis as to how we can heal the world we have made sick.

- There are many minds but only one reality.
- Reality is matter and energy acting in spacetime.
- Minds are intellect and emotion
- Intellect creates science and technology
- Emotion creates religion and art
- Science yields knowledge of reality
- Religion attempts to explain its meaning
- Science gives us the tools to heal and maintain the Biosphere
- Religion must motivate us to use those tools

The world is not ours, we are the world's. We are literally the children of the Biosphere, it is our creator, not in any divine sense, but in the sense that we are products of its evolutionary machinery. To have a future we must heal and care for it using science as a tool but motivated by a deep love and respect for that which gives us life.

AN OUTSIDE OPINION

A team of extraterrestrial ecologists from a highly-advanced civilization would quickly see us for what we are, a newly sapient species, intelligent enough to understand that we are rapidly destroying the health of the Biosphere, but not yet intelligent enough to control our instincts to do so. They would classify us as a Super keystone species. A keystone species being one that significantly alters its habitat thereby impacting many other species, e.g. Beaver, whose dams provide habitat for many other species.   We are a super keystone species because of the magnitude of our impacts. Our exploitation of the Biosphere to meet our needs is simply orders of magnitude greater than its carrying capacity.

Our intelligence is seated in our recently evolved frontal lobes which rest on top of our ancient primitive reptile brain, and the reptile has not given up control. When push comes to shove, we act on instinct, often with great violence. These alien ecologists would know from their own history that progressing from a species with intelligence to an intelligent species, capable of accepting responsibility to heal and preserve its Biosphere, is a rite of passage that all sapient species must undergo or face extinction.

They might upon realizing the degree and rapidity with which we have degraded it, reasonably conclude *the Biosphere is sick and we are the disease*; no doubt amused at the irony that we who claim to be intelligent creatures and who should know better are the *de facto* pathogen. Like a virus, we have changed fundamental processes in our host to satisfy our needs, without regard to its long-term sustainability. The signs and symptoms of this disease include deforestation, desertification, dying oceans, climate change and human caused extinctions.

They would also realize that we instinctively rationalize our destructive behavior towards the Biosphere through our religious myths, which promote the idea that we are "above" the rest of nature. Much of our art, architecture, literature and other endeavors promote myths which perpetuate an incorrect understanding of our place in the Biosphere. This makes changing our behavior very difficult. If these aliens were prone to snap decisions, the cure might be the extermination of our species before we exterminate ourselves and likely take the rest of the Biosphere with us. Intellectually, many of us would agree with the alien's diagnosis if not with its cure.

If we are lucky, these alien biologists, being intelligent, and having seen this evolutionary play before, would simply document that *Homo sapiens*, of planet earth, is an evolving species that represents another chance for intelligence to prove itself a successful evolutionary adaptation, and move on. But what will happen to us if we are left to our own devices? On a future follow up visit, what will these alien ecologists find? A Biosphere restored to health by a fully sapient species or a biologically impoverished world inhabited by roaches, rats and other hearty animals and plants that survived our depredations? I don't pretend to know the answer, but I hope it is the former not the latter. The choice is up to us.

THE TASKS WE FACE

If we are truly programmed by evolution to exploit the Biosphere without restraint how do we stop ourselves and avoid an environmental catastrophe? I would answer that this involves three major tasks.

- We must address the global problem of overpopulation. Universal and equitable birth control by itself will go a long way toward reducing human impacts on the Biosphere.
- We must use our intelligence to solve the technical issues involved with ecosystem restoration on a global scale and develop a science-based plan that is both practical and effective.
- We must motivate enough of the world's population to implement such a plan.

I believe that these tasks are doable but will require the expenditure of much time, money and effort. The sheer magnitude of the effort that will be required is hard to even imagine. The technical problems must be addressed by our best ecologists, engineers and other subject experts. The sociological problems must be addressed by our best social scientists, theologians, politicians and diplomats.

The motivation for this must come from a deep belief that healing the Biosphere is a *moral imperative.* Those who are religious must be led by their respective clergy to believe that it is a Divine Commandment. The clergy of the world's religions must be persuaded through theological or rational arguments to modify their beliefs to allow the use of science as a tool to heal the Biosphere. Those who are not religious must incorporate it into whatever moral code they adhere to.

Technical measures, such as reducing our carbon footprint, will certainly be needed on a grand scale, but as difficult as the technical aspects will be they can probably be overcome given adequate funding and support. It remains to be seen if the same is true of the religious/sociological impediments preventing us from acting. I do not know if we as a species has the collective will to break through our collective bubble of denial and undertake what must be our quest to heal the Biosphere.

We also face the reality that even if we successfully motivate enough of the world's population to act, there is no political infrastructure to carry out ecosystem restoration on a global scale. Not only is such infrastructure missing, there is fierce resistance to having such infrastructure. Patriotism, religion, economics and a host of other differences between states lead to war as often as to cooperation. The United Nations is as close to a planetary government as we have but it is only a loose arrangement of sovereign states not a world state or even a confederation of states. It has been granted by its member states limited authority and only in very restricted areas. It lacks any mandate or capability to implement a planetary ecosystem restoration plan or make and enforce international environmental law.

SCIENCE, RELIGION AND POLITICS

Science and religion are conflicting ways by which we understand the world and our place in it. It has been said "Science gives us means but no ends", meaning that Science is deliberately amoral; which makes it a poor substitute for religion. I believe that the reverse is also true, Religion gives us ends but no means, meaning that Religion is deliberately metaphorical, substituting metaphysics for physics which makes it a poor substitute for science.

Politics is the way we implement our understanding of the world to organize society and constrain human behavior. The struggle between science and religion to dominate politics extends back at least to the ancient Greeks. Beginning about 2500 years ago, Greek Philosophers began to develop concepts of the universe that were based on reason rather than belief. Socrates was one of the first recorded questioners of belief. His idea that one could examine their life through the facility of reason was counter to every tenant of his society because in his time religion dominated politics.

He spent his life questioning accepted beliefs. His statement that "an unexamined life is not worth living" is now a cliché. But at the time it was revolutionary. As often happens questioning accepted beliefs cost him his life.

The idea that one could understand life outside religious dogma began to have a real impact on society at large during the Renaissance and later during European Enlightenment of the 16th and 17th centuries. This led to development of science and the idea that the world was best understood through reason and empiricism. Scientific knowledge can be verified or falsified by anyone with the right training and equipment. Scientific theories are never written in stone. What may seem absolutely true is often only an approximation of a deeper theory. Newtonian physics was only an approximation the deeper understanding of space, time and gravity discovered by Einstein in his theory of general relativity. We know now that general relativity is only an approximation of deeper theory of quantum mechanics. In a real sense, scientific discovery is peeling back the onionskin of reality in search for the central core of how the universe works. This is what makes science unique among all the systems of knowledge.

The gulf between science and religion is widening, as we gain more and more information of how the universe works, this gulf, despite much wishful thinking, cannot be wished away or papered over with platitudes, nor can it be reconciled by suppressing either science or religion. How do we reconcile science and religion? I believe that the best way is to acknowledge that each has a different but vital role in healing the Biosphere.

Some may be reluctant to criticize religion because of a belief in the freedom of religion. But if religious beliefs or any other beliefs are causing us to harm the Biosphere, I believe it I our moral duty to speak out. Not only because these beliefs cause us harm but because they also harm the believers. A religious prohibition against family planning not only harms me and my descendants but the believers themselves because overpopulation is harmful to everyone. But our criticism must be done without malice. We should act as we would toward a friend who was behaving in a self-destructive manner.

Science is grounded in verifiable knowledge, religions are based on beliefs that are simply accepted by the believer. Science accepts as an axiom that the universe arose and operates by discoverable natural laws. No exception to this has ever been found. Each appeal to different and often competing human faculties. Science has increasingly shown light into the dark corners of reality that were the sole purview of religion. Science harnesses our intelligence to know and religion harnesses our emotions to believe. Essentially science is about existence and religion is about essence. Knowing that the Biosphere exists is a matter of science, believing that it has a metaphysical purpose is a matter of religion.

Science provides us with powerful tools with which to reshape the Biosphere to meet our needs. Religion harnesses our need to believe that the world was created and is watched over by a benevolent God. Religion shields us from what our intelligence has figured out, that we are alone in a dangerous scary universe.Religion also provides us a powerful tool for motivating humans to restrain their reshaping the Biosphere to a sustainable level.

To be religious is to believe the unprovable and to be a scientist is to accept the results of experiments and observations. Knowing something is true is very different from simply believing it to be true. Beliefs are not facts to be sensed but rather ideas to be felt, they are not bound by facts and one is free to believe anything the mind can imagine.

One can believe in a God, Goddess or even a pantheon, but their existence cannot be verified or falsified.

Knowledge, in contrast, is bounded by facts which are true whether we want to believe them or not. For example, I know that fire is rapid oxidation which can burn my fingers. This is verifiable knowledge that can be shared with others and independently verified by them regardless of their race or religion. The belief that fire is the product of a fire god cannot be verified and is only true in the mind of the believer.   I can picture God in my mind, but I cannot see God with my eyes.

Beliefs have helped us survive for millennia before there was science. Early on they gave us a comforting explanation for what must have been a terrifying world. This was important in the distant past when we knew very little about how the world worked. Our early ancestors had little factual knowledge of the world and filled in the gaps with magical thinking about gods and goddesses. Imagine living in the Paleolithic period. You are part of a small group of hunter gatherers. Large tribes were not possible because agriculture had not begun to produce surpluses of food. The night terrors must have been like those we see in children in a dark bedroom except that the danger of being eaten by some beast was very real. How better to ward off panic than to take control over your world by tapping into the supernatural forces which seemed to control it.

While we do not know the precise nature of our earliest religions, judging from the archeological evidence and the anthropological knowledge of the ancient beliefs of extant tribes, I believe it is safe to say that early humans could not imagine natural forces great enough to cause the world, therefore it seemed to them that the world must have been formed supernaturally. Believing that supernatural powers are the prime mover in the universe answers a lot of questions as well as calming fears of the unknown. What knowledge they did have was woven into religious myths, astronomical knowledge of the seasons for example served to determine religious holidays as well as set the time of planting. The weather and other processes were explained by reference to the divine. Thunder and lightning are frightening, it is understandable that primitive humans might attribute them to an angry God.

Shared beliefs also contribute to social cohesion. Incorporating injunctions against certain behaviors that were harmful to society were incorporated into religions to enforce them and pass them down to future generations. Incest was likely made taboo following bad outcomes when it occurred. Having it as a divine commandment saved future generations from having to rediscover that it is harmful. Objects of fear may become objects of evil. Such connections reinforce our collective understanding that such things as poisonous snakes and plants are generally to be avoided. In the biblical story in Genesis that it was a snake that led humans to disobey God. It is no accident that the guilty party was a snake and not a cute puppy.

Religious beliefs that originate from ancient myths and superstitions, are at odds with a scientific understanding of the universe. Most of the world's major religions tacitly understand this and have built in doctrines that protect it from scientific facts that contradict beliefs. Religions generally exempt doctrine from the necessity of scientific proof or base beliefs on ancient and incorrect science. To even question doctrine may be heresy. The Roman Catholic Church forced Galileo to recant his observations that the earth orbits the sun and not the other way around and it took centuries for them to acknowledge that he was correct.

The limits of science are the current constraints placed there by the laws of physics. These limits get pushed back as science progresses. Before the telescope and microscope, the limits of our understanding of reality was the acuity of our eyesight and other senses. Modern science has sharpened these through technology, but we still find limits to our knowledge. Physics is constrained by the power that we can apply in particle accelerators, the uncertainties of quantum mechanics and the current state of mathematics. Biology is constrained by the shear complexity of that set of chemical reactions we call "life".

Many religions have beliefs which promote caring for the Biosphere, and many religious leaders including the pope and the Dalai Lama have spoken about taking care of the earth. As a child, I heard sermons on being the stewards of the earth and this doctrine is not confined to Protestant denominations. The problem is that religion is ultimately based on revealed truth and at some point, this truth will conflict with scientific truth.

Science seeks to find the natural laws that govern the universe by investigating its physical, chemical and biological properties. The objectivity of science is the primary reason it is accepted globally and has largely replaced such superstitions as Astrology for the serious study of reality. Astrologers study the same celestial objects as astronomers, but their methods and results are not scientific because the claim that the positions of the stars and planets at the time of one's birth influence his or her personality and fate cannot be verified nor falsified. Astronomy, in contrast, is the scientific study of celestial bodies. Astronomers have amassed a vast amount of knowledge about the universe through careful observation and testing hypotheses. We know that the earth is spherical and orbits the sun and Einstein's theory of gravity explains why it does. Astronomical knowledge is factual whether one chooses to believe it or not, but like all scientific knowledge it is subject to change based on our accumulation of new data.

Unlike human law, natural law is not the product of some cosmic legislature or potentate. Science expresses natural laws as theories, unfortunately the term theory as used by scientists is often misunderstood by those interpret the word to mean an untested idea. That is incorrect, the truth of any scientific hypothesis is determined by its correspondence with reality. As the physicist, Leonard Susskind has pointed out, if a theory contradicts experiments it is the theory that must leave town. Anyone or group with the proper equipment and knowledge can verify a Scientific Theory, or possibly prove it wrong. A scientist's beliefs, nationality, whether they are a man or woman ultimately makes no difference.

Scientific theories typically begin as a hypothesis (basically an educated guess or a mathematical construct as in the case of Einstein's theory of gravity, known as General Relativity). Most hypotheses are rejected based on experiment and observation, but some stand up to testing and are considered theories, at least until a better theory comes along. Newton's Theory of Gravity superseded that of Galileo by accurately predicting the motion of the planets. However, it did not explain certain variations in the orbit of Mercury. Scientists sought for many years to understand this failure. Eventually, Einstein solved this mystery with his theory of gravity which is now the accepted theory of gravity. But even this theory has problems including explaining the effects of quantum mechanics on the physics of subatomic particles. Most physicists believe that a new Theory of Gravity is needed, and much research is being done to discover such a theory.

Science does not depend on belief in the religious sense, no one is expected to believe in biological evolution as gospel, only to understand that it is a verified scientific theory. While the passions of scientists to advocate a hypothesis may to lead arguments, it rarely leads to violence. Beliefs about the use of science on the other hand are something else entirely. Such beliefs are often held in an emotional way much like religious beliefs. The use of nuclear weapons or the need for space exploration. The science behind these things (physics, biology, chemistry etc.) are not beliefs, but the idea that space exploration is part of human destiny or that the use of nuclear weapons is sometimes justified are. The belief that a moon mission would prove that our political system was better than the Soviet Union motivated the United States to spend great resources as did the belief that that the use of an atomic bomb on two cities in Japan during WWII was justified.

This presents us with a conundrum. On the one hand, we can choose to be comforted by the belief that the Biosphere is God's creation and therefore does not need our healing, but in doing so jeopardizes the future of our species. On the other hand, we can accept the frightening fact that the Biosphere is an indifferent natural process and accept our responsibility to heal the damage we have done to it. The conundrum would be resolved if religious believers simply accepted scientific knowledge as divine revelation.

RESPONSIBILITIES OF ENVIRONMENTAL SCIENTISTS

Environmental scientists, myself included, have the responsibility to lead the development of the ecosystem restoration plans needed to heal the Biosphere. They must also play an active role in convincing decision makers to implement them. It has been my experience that many scientists do not wish to become involved in political advocacy, but considering the urgency of our situation, I believe that we no longer have the luxury to remain above the fray. We must be part of a global conversation about the state of the Biosphere and the measures needed to heal it.

But there is an elephant in the room; the science of the Biosphere contradicts widely held religious beliefs about it. We must make it clear that our argument is not with religion *per se.* Our aim should not be to destroy religion, which plays an important role in maintaining social order. Rather our argument is with those beliefs about the Biosphere that are not only factually incorrect but promote activity that harms the Biosphere. While we should respect the right to have religious beliefs about the Biosphere, we must assert the right to criticize such beliefs and actively resist any attempt to have them incorporated into public policy.

The most harmful of these beliefs is that the world was divinely created for us and we are free to do with it as we please. This is stated differently in different religions, but the bottom line is the same. We humans are the raison d'être for creation. This and similar beliefs arose before the advent of science and stem from our instinctive need to avoid the existential angst engendered by a rational understanding that we live in a cold indifferent universe. Our religious beliefs foster the self-delusion that there is a higher power watching over us. In the absence of a scientific understanding humans simply enshrined what they imagined to be this power in religious beliefs and myths.

But I believe that scientific arguments alone will not be enough to change our collective behavior. We must accept that most humans are unlikely to accept the science of the Biosphere outright, regardless of how we present the evidence. Instead, we must enlist the power of religious belief. Scientists must work with religious leaders to translate the scientific necessity of healing the Biosphere to a divine commandment to do so. Healing the Biosphere must become moral issue rather than a purely scientific one. Specifically, believers must be led to believe that a healthy Biosphere is a God given human right and humanities responsibility. This will allow the implementation of ecosystem restoration plans on a global scale, while not requiring believers to abandon their fundamental beliefs about the divine.

The specifics of how religious leaders might interpret the science of healing the Biosphere to be part of God's plan, a commandment of the divine, will vary from religion to religion because of the varying definitions of divinity. The bottom line is that most of humanity will only be motivated to undertake the difficult task of healing the Biosphere if they believe doing so is the will of the deities in whom they believe. The role of scientists will be to provide the scientific knowledge for this task, but it will be up to the world's religious leaders to make the required changes to religious doctrine, scientists cannot do it for them.

As scientists attempt to precipitate what amounts to a religious reformation, we must understand that challenging deeply held religious beliefs is potentially dangerous. The simple fact is that doing science, particularly those aspects of science that challenge religious beliefs about the creation of the universe, is seen as heresy by many religions. Those of us who live in secular democracies are sometimes lulled into thinking that freedom of speech and thought are the natural order of things; they are not. Even a cursory reading history show that the opposite is true, criticizing the prevailing religion for any reason has brought the severest punishments. Nevertheless, I believe that religions must change to the extent necessary to accept that the Biosphere is in grave danger and healing the biosphere should be interpreted as a divine mandate. As scientists we must not shrink from what is our moral duty in providing religious leaders with the scientific knowledge which will allow this to happen.

RESPONSIBILITIES OF RELIGIOUS LEADERS

I understand how deeply religious myths can be imbedded in a person's psyche. I was raised in a fundamentalist protestant church which teaches that the bible is literally true. As a child, I firmly believed that the creation story in Genesis chapter one was a factual account of how the world came to be. Even after I began to doubt this intellectually, based on my growing interest in science, I could not bring myself to stop believing it emotionally. For many years, my childhood beliefs, while gradually fading, hung on to the emotional part of my mind. Eventually I was able to free myself from my years of childhood indoctrination. This experience taught me that one can use his or her intellect to change their beliefs. What we must do is use our intellect to change our beliefs about the Biosphere, which will then change our behavior.

I am not a theologian, but I do know that the world's current religions are amalgams of previous religious beliefs and therefore capable of being changed over time. For those religions, which have leaders who are empowered with divine authority, for example the Mormons, Roman Catholics, some Orthodox Jews and Shia Muslims, the task is somewhat easier simply because theological beliefs and interpretations of sacred texts can be changed to support Biospheric healing by decree. My hope is that the religious leaders of the world, have the wisdom to do what must be done. I believe that it is possible for religious leaders of all faiths to understand the urgency of our worldwide environmental crisis and seek interpretations of their respective scriptures in such a way as to foster the protection and restoration of the earth's natural infrastructure using science as a tool.

This can only be done by believers themselves. Scientists, however desirous of being helpful, can do nothing more than help those religious leaders willing to undertake this task understand what science says about the origin and evolution of life. It is up to the religious leaders to put this into a form that resonates with their theology. Such leaders need to understand that the Biosphere cannot be healed without the use of scientific knowledge which might conflict with religious beliefs. Our job is to convince them to put the changes necessary in a form that will satisfy their members. One way this can be done is to argue that what science has discovered is also divine revelation because it was discovered using our God given intelligence. If this sounds like a scheme to deceive the religious, I would point out that religion is itself a deception as far as its description of the Biosphere is concerned.

RESPONSIBILITIES OF POLITICAL LEADERS

Because the world's political boundaries were drawn with little or no regard for ecological boundaries, most ecosystem restoration projects have been confined within a single national boundary. There are of course exceptions such as treaties protecting migratory birds, fish, whales and those animals harvested for ivory, horn and traditional medicine, but these have more to do with the regulation of hunting and fishing and the reduction of poaching than habitat protection. There have been international agreements protecting multinational ecosystems such as the Gulf of Maine Council, a joint United States and Canadian entity. But these agreements are narrowly defined, and enforcement is up to individual governments who may or may not consider such treaty obligations a high priority.

Because healing the Biosphere will require unprecedented international cooperation, it will likely necessitate rethinking how political boundaries are drawn. Such changes will be difficult to say the least, as they will impinge on our current obsession with national sovereignty, which is the guiding principle of international relations.

The chief barrier to this is that our instincts for social organization have not yet evolved beyond the tribe/clan level. The traits which hold tribes together; strict gender roles, hyper masculinity and femininity, bullying to weed out the weak, compassion for tribe members and hostility to outsiders, strong religious beliefs and taboo's and a hierarchical social structure conflict with the needs of advanced civilizations to maintain large scale social organizations. Healing the Biosphere will require stronger ties between nations. Currently, the nation state is the basic unit of international politics, with voluntary organizations such as the United Nations and international treaties allowing cooperation between nation states. We must move from the idea of national citizenship to one of Biospheric citizenship. One vital step towards this end will be to establish an international agency with the power to enact the rules and regulations necessary to accomplish the overall goal of restoring and maintaining the health of the biosphere.

I realize that it will be difficult to redefine the concept of sovereignty, but regardless of the difficulties a global scale social structure with the authority to enforce its decisions is needed if we are to heal the Biosphere. While this could be done through international treaties it might be better done if we could simply embrace a democratic world government. Whatever form it takes, the world's political leaders have a responsibility to create a working international authority with the power to make the changes that will allow ecosystem restoration on a global scale.

FACING REALITY

Being a scientist, I accept that there is an independent reality discoverable by the methods of science, although to be philosophical honest, I cannot prove it. I once got into an argument with a philosophy professor over the issue of mind versus brain. As a biologist, I believe that the mind is a product of the brain. The professor however argued that while he might be convinced that his brain did not exist it would be impossible for him to doubt the existence of his mind. But he added that he believed in the existence of a reality outside our minds and that only a fool would doubt it. His example was a rattlesnake in the middle of the road. He made the case that only an Agronomist who had taken one course in philosophy would doubt its reality. To make any sense of the world we must accept the existence of a reality that is external to all Minds and is therefore accessible to all minds.

The reality discovered by science is not that of the religious myths and sacred texts we weave to hide it, rather it is what we are left with after we dispense with such imaginings and examine the world around us. But what is that reality? Through science we have learned that the universe was created and is controlled by indifferent natural processes. The concept of God, while good for reigning in our worst impulses is simply unnecessary to explaining the physical universe. For this reason, science does not recognize a divinity or divinities, and none appear as variables in any equation or theory.

We know for certain that the earth is not held up by a giant turtle swimming in an eternal sea, but we are still a long way from understanding the true nature of reality. Perhaps, as some physicists believe, we are simply holographic projections from the boundary of our universe. But ultimate reality does not concern us here. For our purposes, we need only be concerned with what I call proximate reality, four-dimensional space time, the cliff you avoid, and the chair you stumble over in the dark. The reality we cannot deny. This is the reality in which the Biosphere operates and in which it must be healed.

THE FACTS OF LIFE

Although most scientists believe that there is a high probability, even a certainty, that life has evolved on other planets, for all we know now, the earth's Biosphere contains all life in the universe. Even if life is proven to be common in the universe, the immense distances involved between solar systems and the physical constraints on any communication between ourselves and other intelligent species should make us feel a bit like someone stranded on a desert island trying to communicate with other islands by knocking coconuts together. This should give us pause to consider how rare a thing intelligent life must be. Consider that it has taken over 3.5 billion years for the earth to form and intelligent life to evolve on it. Creatures such as us may be like diamonds scattered at random in an enormous matrix of space-time.

But what is life? Most dictionaries define life as the condition or state which distinguishes the living from the nonliving. While adequate for general conversation, this definition raises the question, what is this state or condition? Is life something that exists independent of the physical body and imbues it with Life, perhaps the soul or as the book of Genesis describes it, the breath of life, or is it merely a name for the set of chemical reactions within an organism which arose by natural processes?

While It is true that we do not yet know exactly how life began, the first living organisms evolved from non-living chemical reactions. At this point, we do not understand how this happened, but research will eventually discover the mechanism. It may be that some of these reactions occurred in outer space and the resulting building blocks of life were seeded on earth by comets, but at an early stage in the process simple living cells emerged and from them the process of evolution lead to the formation of the complex global ecosystem we now call the Biosphere.

Up until the late nineteenth century even science assumed that life was a force or principle separate from the physical body. This belief was known as vitalism. Experiments demonstrated that this was not the case and science abandoned vitalism. Scientists now understand that one cannot separate life from living organisms any more than we can separate a wave from the ocean. What we call Life is simply a set of processes that occur in a living organism. Life, as far as we know, only occurs in biological organisms although the jury is out whether and an advanced artificially intelligent entity could be considered alive.

It takes energy to maintain a collection of unruly atoms in the form of a living organism. At death, our bodies begin to break down into its constituent atoms and molecules. This increasing disorder is termed an increase in entropy. Life is essentially an uphill battle against entropy. This difference in entropy is the difference between life and death. Where there is life energy is taken in and structure is maintained. At death, this process stops and decay sets in. Although the Universe is gaining entropy, structures within it including stars and us are temporary events in space-time which persist only as long sufficient energy is available to keep entropy at bay. This tendency to resist the increasing entropy in the universe is apparently built into the fundamental laws of physics. The elements which make up our universe have an amazing ability toward self-organization. Sub atomic particles form atoms, atoms form molecules, molecules are incorporated into tissue, tissue forms organs, bones and connective tissue which in turn becomes a living organism. This process is spontaneous because of the properties of matter itself, without the need for outside (metaphysical) intervention. I believe that it is the driving force behind what we recognize as the process of evolution at the biological level of organization.

But Life is more than an objective process, it is also a subjective experience. Our brain/mind awakens to consciousness in early childhood and we begin to experience the world around us, and that experience, continues until we die. As humans, we experience it with that rare and perhaps unique facility of higher consciousness and language. We see ourselves as the center of the universe. We are so enmeshed in this experience of Life that we rarely think about it. I experience life in all my roles, animal, human, man, husband, father, grandfather, biologist and so forth. This subjective experience of life is not completely conscious in the sense that it is shaped by our inner fears and drives that we may not be aware of. Phobias, for example are just there in our mind, we do not necessarily know how we got them.

From our subjective experience of life that we form our Myths to explain the nature and purpose of the universe. Notions of a creator God or simply nature as a god or goddess become our reality. Our personal myth about life that may or may not correspond well with reality, although a certain recognition of objective reality is mandatory lest we become a tiger's meal or a pathogen's host. Our sense of the beauty of the world and of nature is part of our experience of life. We have music, poetry, stories and ad hoc rules that influence our view of our life and of those around us

But reality never goes away, no matter what we want to believe. If you really want to see our Creator do not bother to consult Sacred Texts or traditions, they contain only the imaginings of those long dead. Instead, seek out a forest or a grassland or any other native ecosystem and you will have found a part of that which created us, the Biosphere or global ecosystem. We and all the other organisms that have ever lived on earth are its progeny. We are literally the children of the Biosphere a creator that is neither a loving God nor Goddess but rather an indifferent natural system in which Life arose and continues to evolve. As far as we know now, we are cogs in the quantum mechanical machine that is the universe. The good news is that this puts our fate in our own hands, for we cannot count on divine intervention or this life being a preliminary to an eternal afterlife. Forget heaven and hell, God and the Devil, all that exists is reality with us as a part of it. We get one shot at life; one chance to heal the Biosphere.

THE EVOLUTION GAME

In science, the validity of a theory is based on how close it comes to explaining the results of experiments and observations. The theory of evolution is such a theory, which simply says that organisms undergo random genetic mutations which either help or hinder survival, helpful traits are passed on and harmful traits are weeded out of the population. First conceived by Darwin and Wallace, and further elucidated by generations of scientists. Essentially, Life survives because species evolve to fit a changing world. Species become extinct because they did not evolve the right adaptations to fit a changing habitat. It is simply the rule of life. I will not attempt to prove that here, that has been well done in thousands of scientific papers and books, nor will I waste time arguing evolution vs. so called creationism or intelligent design; the latter are simply nonsense. Evolution is a settled a theory of science. It can be taken as natural law because it explains the vast body of knowledge that is the science of Biology.

For the non-scientist, evolution can be understood as a game. The object of the game is survival and reproduction. The Biosphere is the arena in which the game is played. There are no spectators. We and all other species are the players. The equipment for the game is the evolutionary adaptations of each player. This equipment is supplied by our genes, which are modified periodically by random mutations. Most mutations are lethal but occasionally a mutation improves a player's chance of survival and is passed down to future generations. There is only one rule, those best suited to their habitat (i.e. the fittest) survive. The referee is natural selection. Winning is temporary because the arena is always changing, and old adaptions no longer fit new conditions. Losing is extinction by not having evolved the right adaptions at the right time. There are no timeouts and the game continues while there are players. But the Biosphere is more than just an arena for the evolution game. It is the game and the game is the Biosphere, one cannot be separated from the other. Without the game, there would be no Biosphere and without the Biosphere there would be no game.

In the larger context evolution is the Biosphere adapting to changing conditions. Species arise and survive or become extinct solely by adaptations, or fitness, for their habitat. Variation in adaptations allows a wider choice of options for genetic selection in the face of environmental change. We see this throughout the biological world. Coloration is one example, while most members of a species are similarly colored because that coloration has been successful, genes for other color variations are always possible either because they already represented in the gene pool or arise through random mutation. Occasionally these outlier genes manifest themselves, almost as a trial run in case the habitat changes and that outlier prove more suited.

Perhaps the most difficult thing for non-biologists to understand about evolution is that there is no plan only a mechanism of chemical and biological processes that emerges from the properties of matter and energy like a flame emerges from a match. This is contrary to everything our instincts tell us. We want there to be a plan, not an indifferent natural process that has no plan. Instead we have genes which are a recipe not a cad drawing. A recipe looks nothing like a cake. You only get the cake if you correctly follow the recipe. A fertilized egg is not a human being any more than an egg is a chicken. A human being is what you get at the end of the process not at the beginning.

Every organism has a built in "drive" to survive. I use quotation marks because this does not have to be a conscious drive, it is more that all organisms have mechanisms for survival. We call these evolutionary adaptations. All organisms have them, plants, animals, bacteria, fungi, you name it will survive if it can. Fangs, claws, drought resistant leaves, long lived spores and innumerable other adaptations have one purpose survival. Adaptations can involve both physical and mental traits; a specialized beak to extract insects from tree bark or the intelligence for abstract reasoning and language. Adaptations arise from random mutations. Most mutations are harmful and are extinguished by natural selection. Successful Adaptations, coded in DNA, are passed on to future generations. Each species, including ours takes its best shot at survival with the adaptations they inherit. Natural selection is about having the right adaptations at the right time. That is what Charles Darwin meant by often misunderstood phrase "survival of the fittest", which is not about physical conditioning as we use the phrase today, rather it is about having evolved the best equipment for a habitat niche.

To survive, an organism must meet its basic needs by employing its evolved adaptations to extract food and other resources from the Biosphere. A predator's fangs help it catch the next meal; the herbivores' hooves help it flee the fangs. Neither predator nor prey is concerned with the future of its habitat. The relationship between predator and prey is a dynamic equilibrium. Without conscious design, this relationship helps insure the future of the habitat. Predators keep herbivore populations in check preventing overgrazing. The predator's population is kept in check by lower birth rates and infant survival in times of low prey populations.

Evolution is incorporated into the fabric of the universe and arises from the properties of that which the universe is made, namely matter and energy. Matter and energy themselves being interchangeable manifestations of some deeper quantum mechanical reality that is not yet fully understood. What we do understand is that our Universe is governed by the fundamental laws of physics and the emergent processes of chemistry and biology. Matter and energy emerged from the big bang or other possible initial event; chemistry and physics emerged from matter and energy, life and biology emerged from chemistry and physics and sentience and sapience arose from biology. The fundamental law of Biology is Evolution, the process that has shaped the development of life from the earliest single cell organisms to the great parade of life forms we see around us.

Despite what we might want to believe, our species is bound by the same rules of evolution as any other organism. Human behavior like that of all animals is linked directly or indirectly to our struggle to meet our basic needs; food, shelter and reproduction. These drives do not depend upon our race or religion or politics, but how we achieve them certainly are. If we are hungry, we seek food, if we are cold, we seek shelter and so forth. All our emotions and drives are evolutionary adaptations for survival. Even our sense of right and wrong is part of our package of evolved traits which allow us to have social order, the most successful evolutionary strategy for humans because we can pool our individual talents. Different societies have diametrically opposed concepts of good and evil because they are struggling to survive under different circumstances. A successful concept of good is one that allows a society to survive another day.

One of my biology teachers maintained that a we are a gene's way of producing another gene. There is some truth to this because our genes existed long before we were born and have the potential to exist long after we are dead. In the larger context, evolution is the struggle of the Biosphere to adapt to changing conditions and thereby preserve life on earth. I am speaking metaphorically because there is no consciousness involved. The Biosphere is an indifferent natural process. There being no indication that the Biosphere is the product of a creative intelligence, divine or otherwise. Creation myths which do not recognize the fact of evolution, while they may seem real to the holder of such beliefs, are someone's fantasy, meant to explain how life began and to give it meaning, thereby protecting us from the truth that the Universe is a cold and indifferent place. While religion is a source of moral teaching and spiritual comfort for believers, religious texts are not science texts books and should not be taken literally.

Most adaptations allow an organism to obtain food, build a nest or survive harsh weather without significantly altering its habitat. Of course, like all things in biology there are exceptions we call keystone species. Beaver are adapted to instinctively dam streams which improves their habitat but also significantly modifies the habitat of many other species. Human intellect allows us such a wide latitude to physically modify habitats so thoroughly and on a global scale. We have, in a few thousand years, become a super keystone species with immense and potentially devastating power. Humans have managed to postpone, at least for the time being, those ecological processes which normally control animal populations. Unlike most other species that are limited to those habitat(s) in which suit them, we can modify most habitats to suit us.

We may explain our actions verbally but the urge to satisfy our needs does not come from anything we may think or say. We have a thin veneer of civilization, if things are going well. In difficult times, as sailors in a lifeboat after the food runs out have demonstrated, we will resort to cannibalism even if we must commit murder. Our violent nature helped insure the survival of one's family or tribe during times of social upheaval. Our ability to love and form social bonds promotes our survival individually and as a group. Our programed behaviors arose by random mutation and because they promoted our survival, were passed down to us; hard-wired in the more primitive parts of our brain.

Our survival instinct and other inborn emotions and drives have been shaped by the evolution game. We do not need a deity to hand down ten commandments to tell us not to commit murder, adultery, or give false witness and so forth, religious injunctions are simply expressions of what we instinctively know. Indiscriminate killing and other destructive behavior is counter to our survival if for no other reason than we could be the victim. Even the evilest persons in history, or serial killers seldom kill indiscriminately, victims are selected because they meet certain criteria, e.g. race, gender, religion, possession of something the murder wants, etc. Even the insane have their own reasons within their paranoid delusions. The tendency to kill everyone other than yourself is simply not a successful evolutionary strategy and is selected out of the population.

But, despite everything I know about the evolution game, I am an optimist. I believe the real problem is not intelligence per se but how we use it. Our ability to plan is a new way to play the evolution game. It gives us the ability to rationally decide not to overburden the Biosphere and to heal it from the effects of our activities.

ECOSYSTEMS, NATIVE AND OTHERWISE

An ecosystem is a human concept, an artificial boundary we draw on a map to delineate a certain area of habitat and the organisms within it, as an aid to understanding how it operates. The areas delineated can vary greatly in scale from the underside of a leaf to the global ecosystem we call the Biosphere. The living and nonliving components of ecosystems are interconnected and interdependent. These connections are as important as the components themselves.

No matter how we draw the boundaries of micro, local or regional ecosystems within the Biosphere they are not separate entities, rather they are parts of a seamless whole. Everything is connected to everything else, directly or indirectly. The Biosphere with its countless interconnections and feedback loops has evolved over billions of years since life first appeared on this planet. Evolution is merciless, the Biosphere we see today represents nature's best effort to adopt to the conditions of an evolving planet.

At any given moment we see a snapshot of an ecosystem, it may look static, but it is not, we are seeing but one frame of a movie. The speed at which the movie plays is determined by many factors and may be very slow over long periods of time. Human activity tends to speed up the movie. A bulldozer, in an afternoon, can destroy an ecosystem that has taken thousands or even millions of years to develop. Of course, nature also has cataclysmic events, earthquakes, volcanoes and the like. The difference is that following natural events such as the eruption Mount St. Helens the slow process of the geologic time scale resume and plant succession begins. Seeds are brought back to the devastated area by wind, birds and other animals. As the vegetation regrows animals return, and the ecosystem is rebuilt.

The area may never return to the way it was before the disaster because of changes in hydrology, topography etc. however an ecosystem will arise adapted to the conditions that now exist. This happened during the last continental glaciation. With the onset of glaciation northern forests moved south, following glacial melting, the process reversed itself and forests returned to the previously glaciated north. In New England, this took thousands of years.

Humans are an integral part of the Biosphere and as much a part of "nature" as any other creature. Never-the-less it is important to distinguish between those ecosystems that have been highly impacted by human activity and those that remain relatively undisturbed. Ecosystems little impacted by human activity systems are solar powered, self-regulating and self-reproducing. Their control systems are also very different from those severely impacted by human activity. Humans manage systems from the top down through simplification and central control. The Biosphere manages itself through biodiversity and chaos from the bottom up. The earth's temperature is not controlled by a simple feedback loop between a furnace and a thermostat. Rather it under the control of thousands of complicated, distributed and interconnected feedback loops at various scales such as the oceanic heat pump we call the Gulf Stream. A centralized control system would be impossibly complex, and the best distributed system is simply the Biosphere itself.

Biologists use the term native ecosystem to refer to ecosystems that have been little disturbed by human activity. Native ecosystems being simply the forest, wetland, prairie etc. that evolved at a given location in response to environmental conditions. The native ecosystem of a city might have been a forest, a desert or any other community of organisms present at that location before it was significantly impacted by human activity.

Native ecosystems are the result of complex interactions between their living and nonliving components. The whole is truly greater than the sum of its parts. Native ecosystems evolve conditions change, they should be pictured as a movie rather than a snapshot. The physical and biological components of Biosphere change over time. Volcanoes erupt, meteors fall to earth, sea level rises and falls even in the absence of human activity. The Biosphere has been able to keep up with these changes over the eons of geologic time. It represents success in the evolution game. Its topography, hydrology, soils and other physical features as well as the biological and physical cycles and processes that occur between the living and nonliving parts of the ecosystem are in dynamic equilibrium. Human activity alters the physical habitat or the biota of native ecosystems, often in unpredictable ways. The more radical the change, the more radical are the impacts.

Native ecosystems are the earth's life support system, a natural infrastructure that makes the earth habitable. For this natural infrastructure to function requires its component ecosystems to be healthy. The key point is that a healthy native ecosystem can produce a suite of valuable functions at one time, such as flood water storage, clean water, aesthetic appeal, and wildlife habitat. This has several ramifications for restoration projects. It means, for example, we that should be very careful when we enhance one function of an ecosystem that the net effect does not reduce other important function to the point where the overall health of the ecosystem is reduced.

Ecosystems are never static even in the absence of human activity. The evolution of an ecosystem depends on many factors from a local to global scale, such as climate and weather, elevation, soil, interconnections to other ecosystems, history of natural disturbances and so forth. They continuously change through evolution and natural disturbances, including fire, insect infestations, blow downs from windstorms, floods, or tree gaps from fallen trees. The evolution of an ecosystem at the local, regional or global scale should be thought of as a movie rather than a snapshot. The opening scene might be some land altering event, such as a glacier, fire or hurricane, frame by frame we see the story of the ecosystem unfold.

The aftermath of this seminal event might be a highly-disturbed landscape possibly down to bare soil. This would be followed by a slow return of vegetation and animals. Wind and birds would bring in seeds from neighboring areas. Those seeds which came from a similar landscape would take hold. There is an order to all of this called primary succession if it begins with bare ground and secondary succession if growing weeds, shrubs, sprouting trees, saplings, small trees and then a mature forest or other habitat depending on the location of the site. Imagine also that this film loops back on itself, replays portions of itself not necessarily in the same order and even reedits itself adding, modifying and deleting scenes. Imagine also that eventually there is another glacier, fire or hurricane. The movie begins again or perhaps there is a sequel. Climate or other changes may cause the forest to be replaced by a desert or swamp. The earth's surface, including its soil, weather, oceans, atmosphere, and much of its geology have shaped and been shaped by the natural processes that occur in the Biosphere.

The distinction between native ecosystems and human dominated ecosystems is important because of the great difference in the rate at which each evolves. Native ecosystems evolve over geologic time meaning thousands or millions of years. Human dominated ecosystems evolve at light speed compared to native ecosystems. The changes brought about by human activity has a deliberateness about it rather than the random way native ecosystems evolve. It is both the speed and deliberateness of human activity that causes so much destruction to native ecosystems. Before European colonization of North America there were millions of passenger pigeons that had been there for millions of years. Natural events, for example, were known to have destroyed thousands of passenger pigeons at one time, but other flocks existed elsewhere which could repopulate an area. When humans, on the other hand, destroyed thousands of passenger pigeons at one location, they would follow the flock and destroy thousands more at another. While this is an oversimplification, the point is that native ecosystems can generally recover from natural disturbance. In fact, some ecosystems require periodic disturbance. Recovery from human disturbance is more problematic.

Human activity can disrupt the very processes which make a native ecosystem possible. Civilization, with its greatly increased population over subsistence cultures and the advent of cities greatly increased the impacts of human activity on native ecosystems. In many cases, particularly in more recent centuries, the native ecosystems of a given location have been virtually eliminated by urbanization, intense agriculture and other human dominated landscapes. Granted, indigenous tribal populations did engage in rudimentary agriculture which included setting fires to remove underbrush, but for the most part their activity had little long-term impact and tended to mimic natural events such as lightning induced wildfires.

The rapidity with which human activity changes habitats particularly affects large animals which evolve at a slower pace due to their slow reproduction rates. In a sense, we are not just another organism we are a pathogen. We are literally making the Biosphere sick, just as any pathogen affects its host. Saying that we are a pathogen has nothing to do with our sense of good and evil, the Biosphere does not make moral judgements, disease organisms are simply playing the evolution game by utilizing their own adaptations and are as much a part of nature as any other organism.

The advent of agriculture that occurred about 10 – 15 thousand years ago is the point in human history when our impacts began to escalate beyond the carrying capacity of native ecosystems. Agriculture allowed for the formation of complex civilizations. Agriculture is perhaps our most successful survival strategy, at least in the short term, geologically speaking. Its long-term success has not been determined. There are indications that it might not be because much of the land farmed in antiquity has been so impacted by agricultural practices that it is no longer suitable for agriculture. The negative impacts of long-term agriculture continue to the present, including depleting ancient aquifers faster than they can recharge and increasing soil salinity on irrigated fields.

A crop field is obviously not a native ecosystem under natural conditions. Rather it is a native ecosystem highly disturbed by human activity. Crop fields depend on human inputs for seed, lime, fertilizer, weed control, etc., and these inputs do not mimic natural conditions. Quite the opposite, human inputs to crop fields are generally intended to create a monoculture of hybrid plants quite unlike anything found in nature. Native ecosystems, on the other hand, furnish their own seed, nutrients, and weed control through complex physical, chemical, and biological processes.

Likewise, a managed forest will have a population of trees of certain species, sizes and ages, that reflect management decisions based on profitability and other considerations. In a "natural" state, the same forest would probably have a very different population of trees that reflected such processes as plant succession, disturbance, and animal activity. The time scale of these two forests would also be different. Human activity usually happens over a relatively short time span, for example, a timber harvest or land clearing for agriculture. Natural processes such as glacial recession and plant succession generally happen over longer periods. There are exceptions such a fires and storms, but even in these instances the presence of humans tends to greatly alter the time scale of natural processes.

One important characteristic that often distinguishes native species from alien species is the tendency for alien species to invade native ecosystems and crowd out native species. Purple loosestrife, a wetland plant introduced from Asia in the last century, tends to crowd out native plants. This is because its natural enemies were left behind. American bison, on the other hand, also came here from Asia but became so integrated into prairies and other North American ecosystems, that it now seems to be the quintessential American species. One difference between purple loosestrife and Bison is that Bison came here between upwards of 800,000 years ago and has had ample time to evolve and adapt to our ecosystems.

In North America, native ecosystems are those that existed before European Colonization. Some of these are ancient and others are young by geologic standards, like those in New Hampshire which only date back to the end of the last ice age approximately 10,000 years ago. Most alterations of native ecosystems have happened following European colonization although Native Americans also altered ecosystems to a minor degree by burning, hunting and farming. In central and South America, complex societies did develop which had significant impacts on native ecosystems.

In New Hampshire, most of the physical environments of its native ecosystems are the result of the continental glaciers, which retreated some 10,000 years ago. Humans, of course, can cause great changes in the physical environment. The grinding action of these glaciers, as well as the great torrents of melt flowing from them as they retreated, shaped and molded the landscape we see around us. Since the glacial retreat, other forces of nature, including at least one period of inundation of much of the state to the Merrimack River by the Atlantic, refined the topography our present landscape. Each of these physical environments supports (or once supported) a characteristic plant and animal community. Each of these physical environments and its associated biotic community is (or was until humans changed it) a native ecosystem.

THE BIOSPHERE IN REALITY

Most of the world's population believes that the world is a divine creation. For them the world is as they want it to be, at least in their minds, but science has found no evidence to support these beliefs. In reality, the Biosphere is simply a complex set of physical/chemical/biological processes. I believe that the tendency to believe otherwise is an evolutionary adaptation and therefore it is pointless to use scientific arguments to change people's minds in this regard. Our approach to enlisting believers in the effort to heal the Biosphere must be through the world's religious leaders who must be persuaded to incorporate into existing religious dogma the idea that healing the Biosphere is God's will. I will discuss this in more detail later. For now, I want to explain the scientific understanding of the creation and operation of the biosphere to those who are willing to accept such an explanation.

If we are to scientifically heal the Biosphere, we must, at least temporarily, put aside all our romanticized notions of it. This is the Biosphere stripped naked of all imaginary elements including God and other imaginary beings. In this Biosphere Mother Nature and Gaia are only metaphors and God is real only in the imagination of believers. The scientific explanation of how the Universe came into being does not require a belief in anything beyond the laws of physics. For the religious believer, the Biosphere functions because it is part of God's plan. For the science minded the universe essentially creates itself, being an emergent property of matter and energy acting in spacetime. Another way of putting it is that for religion, intelligence (in the form of God) preexisted the universe and for science, intelligence (in the form of intelligent organisms) arises from the universe by the process of evolution.

With minor exceptions, the sun powers the Biosphere. The physical, chemical, or biological components and processes of the ecosystems that make up the Biosphere support every aspect of our lives. Green plants use the sun's energy to produce sugar, water and oxygen from carbon dioxide in the air. The cycling of oxygen, nitrogen and carbon dioxide by plants maintains a balance of atmospheric gasses suitable for life. Pea plants and other legumes have bacteria in their roots which convert atmospheric nitrogen into the building blocks of proteins. The sugar produced provides energy for herbivores who in turn furnish energy for carnivores. The decomposition of dead plants and animals by microorganisms recycles nutrients back into the Biosphere.

Our species and all other species arose and evolved in the Biosphere. Indeed, life on earth is inseparable from the Biosphere's ecosystems, but the Biosphere is a harsh and unforgiving place and no amount of wishful thinking will change that fact. Realizing this for the first time can be a bit disturbing. It is not the Biosphere most of us would prefer. Given a choice between an indifferent universe and one in which a caring God watches over me, I personally would choose the latter. That choice would be obvious for most people. It is like the choice between being tucked in at night by a loving parent and being thrown out into the street to fend for yourself. We want to believe that we are special creations and that our needs have priority above those of other animals. We tell ourselves Creation myths to hide the truth from ourselves. We deny reality and convince ourselves that we are no longer alone but part of a divine plan. The problem is that this fiction promotes the destruction of the Biosphere. Humanity must accept that we have only one Biosphere, this one, even if it is not the one we would prefer.

THE BIOSPHERE IN OUR MINDS

In the last chapter, I discussed the Biosphere as science describes it, existing in an objective reality outside our mind. But what exactly is the mind? I know that I have a mind, to doubt it would be absurd for the simple reason that the mechanism with which I would doubt my mind is my mind. Descartes, the seventeenth century philosopher and mathematician who sought to find the most basic belief on which to build a philosophy, said "I think therefore I am". This is about as succinct a definition of mind as one would want. Our mind is what we think with, the seat of our consciousness. But it is also the seat of our unconscious, those primitive emotions and drives we have inherited from our forbearers on the evolutionary tree.

Our intelligence is seated in the most recently evolved portion of our brains, the cerebral cortex. The cortex is built on top of the more primitive portions of the brain we inherited from our reptile ancestors. Our thinking brain is therefore not independent of our more primitive emotional brain. Much of what happens in our mind is not accessible to our sentient self. Our basic drives, emotions, fears and so forth are more felt than thought. Our attempts at rational thought are influenced by our more basic drives. Our behavior as a species is driven by our instincts and not by our intelligence.

The mind is ethereal in that we distinguish between the mind and the brain, some go so far as to posit our mind as being separate from our body. I do not go that far, rather I believe that the mind is an emergent prosperity of the brain. By this I mean that the brain, as a complex biological organ produces the subjective experience, we call our mind. This is evidenced by the fact that every mind appears dependent on a living brain; dead brains appear to lack a mind. It could be argued that the mind or soul of a dead brain, at least a human brain, has left the body and still exists somewhere, but this cannot be proven one way or the other.

For the purposes of this book we need only think of our mind as the receiver and processor of vast amounts of sensory input, visual images, smells, tactile sensations etc. Our sanity demands that we have a coherent mental model of the world in which sensory input has a coherent meaning. Otherwise we would be left with mental chaos of disjointed images like we experience on a carnival ride.

Our mind forms, over our lifetime, a mental model of the world cobbled together from sensory input interpreted by our education, emotions, instincts, thoughts and other mental processes. This model is much more than a simple picture of the outside world. I liken it to a painting rather than a photograph. Our model is never completely accurate because senses perceive the world in a very limited way. We are only aware of three dimensions of space and one of time, although string theory suggests there may be as many as ten dimensions of space. Our eyes see only a limited spectrum of light and our ears a limited frequency range of sound and we cannot directly see an atom. Our model of the world is a simplified version of reality. Never-the-less, for all practical purposes, this model is the Biosphere we interact with. The state of our internal model of the Biosphere governs is how we "see", value and act toward the real Biosphere.

We create this model instinctively beginning with our earliest childhood sensations and it is never complete. Every sensation is interpreted and affects the state of the model. We do this like we breath, unconsciously until we think about it. Fig. 1 is a simplistic schematic of how this model might work. Our internal model of the Biosphere can be thought of as having a set of scales. On one side is our knowledge of reality, objects and processes that can be tested using the methods of science. On the other side are beliefs which cannot be tested. The state of this model at any time is dependent upon many variables which correspond to the commonly understood aspects of Mind including our drives, intelligence, sentience, sapience, values and personality to name a few. It is influenced by our genetic makeup and the culture and religion in which we were raised. Our individual models of reality are unique and contain our accumulated beliefs and knowledge about the world.

Our individual models of reality are unique to us because our set of life experiences is unique. Scientists like myself tend to believe that the universe is an evolving process governed at the most basic level by the laws of particle physics and general relativity. Those who are religiously inclined believe that the origin and operation of the universe is a product of divine will. My mental model, for example, has been informed by both my early religious experiences and my training as a biologist.

The state of our mental model of the Biosphere is not constrained by reality. One of the beauties of the mind is its freedom from the laws of physics. In our mind, we make the rules. Our thoughts are limited only by our imagination. We can create complete worlds of fantasy and we can also believe fantastical things about this world. But as I will explain, our ability to do this is both a blessing and a curse. The knowledge that fire is hot, may exist side by side with the belief that there is a fire god who makes fire hot. An individual is free to include fairies and other mythical beings in their model and do it in a way that they believe that they are real. I would add that I do not mean we should not apply our imagination and artistic creativity to our understanding of the Biosphere. We can keep God in our model if doing so comforts us. We can compose music and write poetry about the wonders of nature if we are so inclined. But in doing so we must not lose sight of the fact that any characteristics we add to our internal model of the Biosphere that are not based on reality are metaphor not reality.

Our mental model of the Biosphere is important to the health of the Biosphere because it determines our actions toward it. Our problems arise when we begin to treat the Biosphere it as it is in our imagination, rather than as it really is. This failure was understandable, in the pre-science era before we understood how to separate our myths about reality and reality itself. But now we have science to clearly reveal how the real Biosphere works and this is the Biosphere we must heal.

Our internal model of the Biosphere is not composed entirely of rational propositions. If that were so we would be like Spock of the Star Trek universe. All of us incorporate fantasies into our model which give our lives meaning and purpose. This is part of our inborn tendency to deny the fact that the Biosphere is a harsh and dangerous place. A cloud becomes an elephant, a sunset a thing of beauty and the Biosphere is personified by Mother Nature. These are our imaginative interpretations of purely physical phenomena. The balance between fantasy and reality varies greatly from person to person. For some fantasy outweighs reality and their mental model of the Biosphere bears little resemblance to reality. Regardless of the accuracy, our model of reality, it is for all intents and purposes our reality. It is this internal reality that determines our behavior toward the real Biosphere. While this may bring about positive behavior, certain fantasies can engender great harm to the Biosphere. Especially religious fantasies that lead us to believe the world was created especially for us and we are free to do with it as we wish.

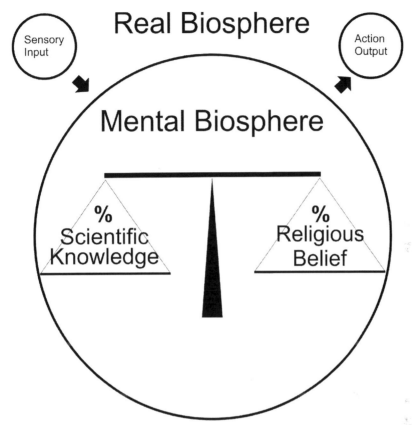

Fig. 1 Our mental model of the Biosphere is weighted toward scientific reality or religious fantasy.

Despite of the fact that the scientific model of the Biosphere is factually correct, most of the world's population clings to a mental model of the world in which the Biosphere is a divine creation. This divine myth varies from culture to culture but is for all practical purpose's reality for believers. These divine myths about the world are based on beliefs going back thousands of years and are simply incorrect. They perpetuate a false picture of reality. Sacred Texts are moral teachings not text books of Cosmology and Biology. They are wrong in their portrayal of the Biosphere and our place in it. Our creation myths simply do not distinguish between fantasy and reality. More importantly, they are wrong in ways that promote and excuse our ongoing destruction of the Biosphere.

While theological details differ, the Hebrew/Christian Bible and other sacred texts and traditions tell us that the world is the result of divine will. It was created by a deity that operates in a metaphysical realm outside the physical reality we see around us. These divine entities are free to violate the laws of physics by miraculous means without incurring any of the consequences expected because of the laws of thermodynamics. For the believer, the Biosphere is a divine creation; a mystery to be accepted and believed not an unknown to be investigated. Religion is less interested in how the world works physically than in understanding the will of the Creator.

Most sacred texts are short on detail about the physics and biology of creation, Genesis boils the process to seven days, and God simply creates Adam out of the dust of the earth, without any indication of how the details of human anatomy came to be. On the other hand, sacred texts and traditions generally go into detail about leading a moral life, the book of Leviticus being an example. In this sense, religion focuses on the moral aspects of creation rather than on the physical processes underlying it. Science, takes a very different approach, focusing on the physical and biological processes of creation without making moral judgements about it.

The more science discovers about the universe the less likely it becomes that the gods described in the world's major religions exist. That is not to rule out the possibility that some Deist divinity exists who set the whole shebang in motion and then let it run by relativistic/quantum mechanical rules. But the likelihood of even this form of divinity is fading rapidly. Intellectually it is clear that the active, intrusive gods of sacred texts are figments of our imagination created to allay our fears of being alone in scary indifferent universe.

WHY WE ARE DESTROYING THE BIOSPHERE

It is obvious to me that Despite this shortcoming, we must play the evolution game, but must play the evolution game, using our evolutionary adaptations of intelligence coupled with our superior manipulative abilities, to modify and exploit the Biosphere to meet our needs, but at the cost of destroying the very global ecosystem that sustains us.

Our destructive behavior toward the Biosphere has been attributed to a whole spectrum of causes from original sin to greed driven capitalism, but I believe that these commonly stated reasons are more explanations or justifications for our behavior than the cause. It is my hypothesis, based on many years' experience in ecosystem restoration, that we act as we do because we have not evolved to the point at which intelligence rather than instinct controls our collective behavior toward the Biosphere. Our actions are not sinful or caused by any defect, they simply reflect how the evolution game is played. We are driven by our instinct to exploit our habitat the same as any other species, our intelligence functions primarily as a tool to do this. In the short term, on a geological scale, intelligence has benefited us, but its benefit in the long term is an open question. We have named ourselves *Homo sapiens*, but we are not so sapient when it comes to planning for the long-term future of our species.

Our intelligence gives us the power to modify our habitat to meet our needs. Other keystone species share this ability to a very limited extent. But most species must survive (or fail) in their habitat as they find it. This ability has lulled us into thinking we are exempt from the laws of natural succession, but in the long term it has the opposite effect. We are rapidly making the Biosphere less suitable for our survival by exploiting it beyond its carrying capacity. We are more susceptible to natural selection because our destructive behavior is making us less fit for our habitat by making our habitat less fit for us.

We have not used our intelligence to institute reasonable limits on population, habitat modification, biodiversity reduction and a host of other ecological concerns on a global scale. Instead we just dumbly forged ahead tearing away at the fabric of the Biosphere to meet our short-term needs. We paid no more heed to the future than yeast making beer. Eventually the sugar runs out and their excreted waste kills them. Yeast can be forgiven for their lack of insight because they do not have brains, but we do and should know better. Perhaps we should rename our species *Homo semi-sapiens.*

Before we began to farm, human populations were small and scattered and our activities had little impact on the Biosphere. Once agriculture became widespread after the last ice age roughly 10,000 years ago, human populations and their impacts on natural ecosystems increased dramatically. Agriculture has several direct and indirect impacts on the Biosphere. It provided surplus food allowing populations to grow and by freeing many people from the daily necessity of gathering or hunting food, allowed the development of complex societies and advanced technologies increasing the demand for more land to be converted to agriculture. There were also direct impacts on native ecosystems from common agricultural practices including but not limited to forest clearing, grazing, draining, burning, plowing, tilling, irrigating, and planting monocultures of hybrid plants.

As technological development accelerated, and we became caught in a positive feedback loop of increasing population requiring increasing exploitation of the Biosphere, which increased population. Each time we used technology to solve an environmental problem we created other unanticipated environmental problems. Essentially, every environmental problem is the direct or indirect result of the solution to a previous environmental problem. Our use of technology in this manner, while very effective in the short term, began to cause more long-term problems than it solved. Discoveries that seemed totally benign, had unexpected consequences.

For example, fire kept us warm but large-scale acquisition of fuel, either through deforestation, mining or petroleum extraction degraded ecosystems.

Agriculture freed us from the dependence on hunting and gathering but allowed our populations to grow beyond the carrying capacity of our habitats. Improved hunting and fishing equipment and methods reduced or eliminated some wildlife populations. These impacts were small at first but grew like a snow ball rolling downhill. Currently, the damage we have done to the Biosphere threatens our global life support system. The list is very long and well documented; climate change, desertification, over population, unsustainable agriculture, to name a few. Whole ecosystems, such as the American Tall Grass Prairie and the bottom land hardwood forests of the Mississippi Alluvial valley have been lost or exist only as isolated remnants.

Many would argue that the development of high culture proves we use our intelligence for higher purposes than simple survival in the evolution game. I believe to the contrary, that even high culture serves the same evolutionary purpose as a spear in the jungle. Religion, art, music and all those activities we consider examples of the human spirit are almost universally aimed at promoting and justifying what we do instinctively and that is to take what we need for survival. For example, martial and patriotic music and art helps maintain social cohesion, particularly during times of war. Religious art and music obviously promote religion which also helps maintain social cohesion. I admit that there are instances in which reason wins over instinct and we pass laws and otherwise attempt to protect elements of the Biosphere. But such cases are rare in the scheme of things and environmental laws are often poorly enforced and easily circumvented.

I believe that it would be sad that after all the billions of years and countless evolutionary steps necessary to produce us that we should, in an eye blink of geologic time, cause our own extinction. But sad or not our extinction is a real possibility given the serious damage we continue to inflict on the Biosphere.

OUR BUBBLE OF DENIAL

Religious beliefs about the Biosphere are beliefs detached from the real Biosphere. Such beliefs make what is imaginary real, and once one steps into an imaginary world anything is possible. The Biosphere, as a complex natural process on a lonely planet in a vast universe, disappears and in its place, is a simplified world divinely created and managed at the center of a small universe surrounded by the metaphysical realm of God. Our imagined universe becomes for us, a bubble of denial.

I think that we created religion to explain the unexplainable and to enshrine our denial of harsh reality in a form that is resistant to rational argument. From the evolutionary point of view, belief in a metaphysical deity helps us overcome the existential angst engendered by the dangers of the world. We simply have an inborn need to believe that the world is about us and that the universe somehow cares about us. This need is built into our psyche by evolution, because believing improved one chances of surviving over those who did not.

Despite what we would like to believe, we live in the real Biosphere. We are players in the evolution game and like all other species we are driven by the instinct to survive. We use our adaptations to survive in the here and now, not for our long-term survival. But our level of sentience is a two-edged sword, it not only facilitates our instinct to survive, it allows us to see the Biosphere as it is, a collection of indifferent natural processes governed by the laws of physics and chemistry. This knowledge has the potential to cause existential angst which would lower our chances for survival. As a counterbalance we have evolved a denial mechanism that blinds us to what our intelligence makes clear. We essentially live in a bubble of denial, within which we substitute a benign world created and managed for us by a deity of our own creation.

Sacred myths also excuse us from the responsibility of living within the carrying capacity of the Biosphere. Particularly religious myths which promote overpopulation. Our myths free us from responsibility for the health of the Biosphere because it is part of some divine plan and therefore out of our hands. They are not the cause of our behavior towards the Biosphere rather they are our way of justifying it.

This must change, or we risk self-extinction. I believe that our only option is to change the way we play the evolution game. We must accept the Biosphere as it is not as we would wish it to be. Our actions toward the Biosphere must be guided by our intelligence not our instincts. Scientists and others understand this, but most of humanity does not, simply because humans need a sense of hope and it is hard, but not impossible, to have hope realizing that the world is not about us.

Denial is so deeply imbedded in our unconscious as to make otherwise intelligent individuals believe fanciful creation myths even in the 21st Century. Because our denial is instinctive, no amount of rational argument will have more than a temporary impact on humanities collective behavior.

If most of the world's population is unable to play the evolution game with intelligence rather than instinct, those of us who understand what is at stake must help translate what is a scientific necessity into a moral imperative. This is true if for no other reason than the chaos which occurs when a civilization depletes its resources to the point of famine and insufficient water supply. I will discuss how this might be done later, but for now I will simply say that the goal is to convince religious leaders that what must be done scientifically to heal the Biosphere must be communicated to their followers as divine commandments.

VALUING THE BIOSPHERE

Like all species, we are driven by instinct to obtain sustenance, shelter and an opportunity to reproduce. But humans are not, like some species, solitary creatures. We exist in organized societies and these societies constrain what individuals are permitted to do to survive in the Evolution game. These constraints, that is what behaviors are permitted and what are prohibited, are expressed in the shared values of a society. I believe that the value systems of most societies whether preliterate tribes or first world nation states are ultimately based on the predominate religion of that society. While they may be codified in civil law, they most often derive their authority from divine commandments found in oral myths or sacred texts. This sense that the values of a society are part of the divine order of the universe are the foundation on which a societies social order rests. Even those individuals who are not particularly religious are forced by social norms to conform or risk being ostracized or worse.

Organizing societies around values accepted as divine commandments has been a successful evolutionary strategy throughout human history. We form religions to explain and give divine authority to our values, but it is important to understand that the religions themselves are not the source of our values. Our values, or more specifically, our ability to form values is simply another of our evolutionary adaptations for survival. Values such as good and evil are universal, but exactly what is deemed good and evil varies from society to society because the environments in which societies must adapt vary. Desert dwellers developed different social values around water use than societies that originated in humid regions where water scarcity is not an issue. The bottom line is that value system of every society represents the set of values that have helped it survive.

But what does this have to do with healing the Biosphere? The answer is simple, how a society values the Biosphere and its component ecosystems has major impact on how that society treats it. For simplicities sake I will discuss in this chapter three ways that society can value the Biosphere and the implications each has for its health.

The first way is to value the Biosphere is for the resources we can extract from it, e.g. timber, peat, minerals etc. without regard for the impacts of extracting these resource extraction on its health. Essentially this is the economic value of an ecosystem. The problem is that economics, which is the study of the allocation of scarce resources, completely misses the point when it comes to valuing ecosystems. Economists do not have a good way to put a monetary value on a whole functioning ecosystem, only on its constituent parts extracted and sold on the open market as well as the market value of the land on which the ecosystem exists converted to other uses. Unfortunately, there is a powerful lobby of miners, loggers, land developers and others who see this as perfectly acceptable. This is widely expressed as the "wisdom of the market place."

It was demonstrated many times to me over the course of my career that idea that economic value always trumps such values as a love for nature or a sense of beauty in a pristine landscape. Economic value is almost universally accepted as an unquestioned goal of public policy. The very infrastructure of the United States and most of the world was built on this principle. Roads were built through wetlands with little thought as to how they disrupted adjoining ecosystems. Such projects are rarely abandoned because of environmental concerns. At best, the worst impacts may be mitigated by some feature such as a fish ladder built into a dam. But even in such cases the economic value of the fishery may play as large a part in gaining mitigation as the intrinsic value of an unrestricted river.

Economic value is expressed in units of money. The scarcity of gold makes it worth more dollars per ounce than lead. Obviously, this same principle applies to the components of an ecosystem. Scarce marketable resources extracted from an ecosystem gain in value from their scarcity. But there is no way of measuring the ecological value of a native ecosystem in terms of its contribution to long-term human survival. The ecological value of a rhinoceros' horn is highest when it is attached to a living rhinoceros, its economic value on the other hand is highest when it is ground up as traditional medicine. Likewise, the economic value of a tree is highest when it is cut down and sawed into lumber, but its highest ecological value is when it is growing and producing oxygen in a forest ecosystem.

The value of an intact native ecosystem often has an inverse relationship to scarcity. The extraction of resources itself will almost certainly reduce its ability to function. Scarcity of certain ecosystems, such as individual prairie pothole wetlands, become less valuable as they become scarce because it takes many of them of varying wetness to function as a complex in providing habitat for waterfowl. Removing individual wetlands in a complex by the draining for agriculture and development does not make the remaining wetlands more valuable. The scarcity of prairie pothole wetlands caused makes the remaining wetlands less valuable for ducks. Placing economic value on the components of an ecosystem may result in its loss rather than its protection. Doing this on a global scale has greatly reduced the ability of the Biosphere to support life on this planet.

It may be difficult for many to accept, but it is obvious that the idea that God gave us the world to subdue is really a justification for valuing native ecosystems for their extractable resources rather than as intact functioning ecosystems. This idea of ownership of the earth is not confined to one religion. With few exceptions it is accepted worldwide and has likely been part of religions since they first formed. As a result, most humans, most of the time, value the Biosphere for its "natural resources" rather than as collections of functioning ecosystems. The religious underpinning of this belief makes it immune to rational argument to the contrary. I know because I have tried many times in many meetings.

The resource value of native ecosystems is promoted in the bible and other sacred texts, in the myth that we own the Biosphere by divine right, rather than being a part of it. An example is the creation myths in the Book of Genesis, which form a central part of our Judeo/Christian heritage; the first book of the Christian/Hebrew Bible. In it, God creates the universe and us in seven days. Early on, God giving man a special place in the scheme of things.

- Genesis 1:26 And God said, let us make man in our image, after our likeness: and let them have dominion over the fish of the sea, and over the fowl of the air, and over the cattle, and over all the earth, and over every creeping thing that creepeth upon the earth.
- 1:27 So God created man in his own image, in the image of God created he him; male and female created he them.
- 1:28 And God blessed them, and God said unto them, be fruitful, and multiply, and replenish the earth, and subdue it: and have dominion over the fish of the sea, and over the fowl of the air, and over every living thing that moveth upon the earth.

Variations of this belief are common to most of the world's major religions. Regardless of whether this myth is literally true, acting as if it true has been a receipt for disaster as far as the Biosphere is concerned. You might ask "what was God thinking?", to give the world to such an irresponsible species. Taken literally we have *Carte Blanch* to do with the earth as we see fit. While the text may suffer from translation, the term subdue is clear.

The second way of valuing native ecosystems is for the beneficial functions they perform, including, but not limited to fish and wildlife habitat, flood storage, recreational and educational opportunity, producing forest products, supplying oxygen, and improving water quality. At a global level, natural ecosystems, particularly forests, are important in maintaining the earth's temperature and controlling carbon dioxide levels. Scientists believe that native ecosystems are vital to the long-term support of life on earth; in ways, we are only beginning to understand.

While it is possible to enhance one or more of an ecosystem's functions without harming its overall health, it must be done with caution. A native ecosystem functions best when intact and free of anthropogenic stressors. For our purpose we must look beyond any individual function it performs and consider the whole ecosystem. We must make restoring the health of native ecosystems the goal rather than optimizing one function at the expense of the others. We should not become the hydrologist who only considers the flood water storage capacity of a wetland, or the rancher who only considers the grazing capacity of a short grass prairie, the forester who sees only standing timber or the biologist who values fish and wildlife only in relation to hunting and fishing.

I was involved in the development of a national method of evaluating wetlands for various functions such as wildlife habitat, flood damage reductions and so forth. I was asked to do this based on my previous experience developing such a method for the states of Connecticut and New Hampshire. The problem is that when scientists try to persuade a landowner about the importance of a wetland and why it should not be filled and paved over for a parking lot, they can only refer to the functional values of the wetland, e.g. flood control, wildlife habitat etc. Unfortunately, functional values were often trumped a landowner's desire to profit from the conversion of the wetland to other uses.

The third and most important way of valuing a native ecosystem is intrinsically, which is hard to define precisely but it has to do with the strong emotional attachment some feel for native ecosystems, regardless of any economic consideration. This can be expressed, sometimes pejoratively, as being a "nature lover" or "tree hugger". Intrinsic value is considered a higher value than a functional value in the conventional hierarchy of values. An intrinsic value is a deeply held belief that a thing is important in and of itself over and above its ability to perform specific functions. It a recognition that the sum of the processes that make up an ecosystem is greater than its parts. Intrinsic value takes us to a place beyond knowledge and into the realm of belief, where an ecosystem engenders love in those who understand it. Most wetland scientists may give rational reasons why they are working in the field, but it has been my experience that deep down they simply love wetlands. They felt as I do that, we have a moral imperative to protect the natural world that stems from this love of it. Indeed, there was an unspoken belief that being an environmental professional was a calling rather than just a job.

Americans generally love the statue of liberty intrinsically as a symbol of what we hold dear, and only secondarily for its functional value as harbor beacon or for its value as scrap bronze. In a similar way an individual may know the functional value of a native ecosystem but he or she loves it for its intrinsic value. It is my experience that this is what motivates professionals and volunteers in the environmental movement. We simply love wetlands and other native ecosystems, although we professionals would never say that at a public meeting for fear of sounding sentimental. I believe that this deep love of ecosystems, the Biosphere, nature, or however it is expressed is a form of religious belief, it imbues native ecosystems, at least metaphorically, with a value beyond their physical existence. It is this love that must be translated in to the language and doctrines of the world's religions, to provide the moral imperative for healing the Biosphere.

Since healing the Biosphere requires a long-term view, one of the things that must change is how we value it. Because our values stem from our instincts not from our intellect they are hard to change. Most of the population cannot change until we evolve into a species with a level of intelligence great enough to adopt a new strategy in the Evolution Game in which our intellectual understanding of the long-term value of natural ecosystems to our survival as a species is stronger than our instinct to exploit them for short term gain. Unfortunately, we do not have the luxury of time to wait for any such fundamental change in human nature. Modern humans are what they are, at least for the foreseeable future.

The arguments for ecosystem restoration that I made as a resource professional were constrained by public policy. I could not say for example, that ecosystem restoration was an intrinsic good. I was constrained to argue the functional values of native ecosystems. Those on the other side of the argument who argue for the unrestrained exploitation of natural resources can invoke the unquestioned belief that economic development is the highest intrinsic value of our Capitalist economic system. This belief in the rightness of an economic system is as un-testable as any religious belief and as closely held. Such beliefs are impervious to rational argument very often trump any belief in the intrinsic value of wetlands.

Certainly, Christians for example can want to take care of the Biosphere the idea that we have been made stewards of the earth by God is a great motivation for some believers. Why are such sentiments not enough? The reason in my opinion is that the earth of Christian stewardship is not the real world, it is the earth of myth, the earth that is part of a divine plan. It is not the earth that is an object in space-time subject only to the laws of physics. The created world of Genesis accepted by Judaism, Christianity and Islam for example is not even a good metaphor for the real world. The real world is much more than ball of created stuff populated by the beasts of the fields and the birds of the air. It is the substrate for the Biosphere, in which living organisms wage an uphill fight against entropy, a complex game of evolution where surviving is winning and losing is extinction.

I do not minimize the efforts of individuals, organizations and governments which have tried to apply reason to environmental problems. I know from personal experience that there are many people dedicated to protecting the Biosphere. What I am talking about is our collective behavior. Concern for the impacts of on the Biosphere for the majority of human is an afterthought at best. Attempts to restrain ourselves through environmental laws, treaties, activism etc. have made little headway in the face of nationalism, war, economic development, population increases and a host of other emotionally driven activities. We are simply not governed by reason when it comes to protecting the environment. We continue to exploit the Biosphere as if it was an unlimited resource rather than as a functioning ecosystem having a large but finite carrying capacity.

We must ask ourselves whether we are condemned to destroy the Biosphere because valuing ecosystems for the resources that can be extracted from them is instinctive? How do we overcome our inborn strategy in the Evolution Game? As I have tried to make clear we cannot do this by intellectual argument directly. Our intellect is simply not in control. What I suggest may be hard to understand at first glance but bear with me. I am suggesting that we use our intellect to shape our beliefs and consequently our sense of value. I think that the only way to do this is through the power of religious belief. I believe that this is possible because of a reciprocal relationship between our instincts and our beliefs. While our beliefs and their implied values come from our instinct to act in the short term, I believe that the reverse can happen, our beliefs can, under the right circumstances override our instincts. We see this happen every time some heroic individual sacrifices themselves to save a stranger's life. The belief that this is the right thing to do overrides our instinct to not run into a burning building to save a stranger. This happens in an instant without rational thought. Reason only comes into play when we were being taught the belief that it was a moral imperative to save lives even at the cost of our own. This idea is expressed in Christianity by the words of Jesus to the effect that the greatest love a person can have is give one's life for a friend.

My suggestion is this, those of us who understand our situation must attempt to change religious beliefs in such a way that a healthy Biosphere becomes a human right and humanities responsibility. In other words, healing the Biosphere is a moral imperative by divine commandment like the prohibition against murder. I have alluded to this in an earlier chapter and I will say more about how this might be done later, but the point here is that creating such a belief on a wide enough scale will fundamentally change the way humans value ecosystems. I realize that this is will be a difficult if not impossible task but believe that putting the power of divine commandment behind the idea of a healthy Biosphere is the only way to change human behavior on a large enough scale to make a difference.

A RELIGIOUS MORAL IMPERATIVE

What is a moral imperative? For our purposes, I define it as a cultural or personal norm of proper behavior. More than that moral imperatives call us to act, either mentally (e.g. prayer) or physically (e.g. build a cathedral). I believe that humans have an instinctive drive to create such norms, making anarchy rare and temporary. Moral imperatives rest upon an accepted moral authority. For most societies, moral authority is vested directly or indirectly in its majority religion. This is true even in the United States which is nominally a secular democracy, but which is largely based on the moral precepts of the Judeo/Christian tradition. Religions claim their moral authority as the interpreter of Divine will, whether it is the father god of the Abrahamic religions or a divine but impersonal force such as the Tao or Dharma. The details vary from religion to religion, but each considers its moral imperatives to be of divine or at least metaphysical origin.

Moral imperatives create order and purpose and are present in all established societies, whether a tribe, clan or nation state. They may be passed down as oral tradition or in more advanced societies encoded into civil and/or religious law. Nothing of significance happens in a society without being heavily influenced by them. To resist or violate them always has consequences, sometimes severe.

The Judeo/Christian tradition provides the most widely accepted set of moral values that underlie society in the United States. We cite the ten commandments as the bedrock of civil law, in court I would swear on the bible to tell the truth, the whole truth and nothing but the truth so help me God. Likewise, other cultures root their values in their respective understanding of god. I realize that this is an over simplification, especially in a modern multi-cultural society but my point is that for most people, the ultimate source of values is some form of the divine, whether it is valuing human life, being truthful, or being honest etc. And because a societies secular law is derived or at least strongly influenced by that societies' dominant religious values, non-believers are also forced to recognize them.

The ten commandments of the Torah/Old Testament are a good example of elevating basic rules of conduct to the level divine commandment. Notice the first four commandments define the source of moral authority while the last six are the rules of conduct being elevated.
1. You shall have no other gods before Me.
2. You shall make no idols.
3. You shall not take the name of the Lord your God in vain.
4. Keep the Sabbath day holy.
5. Honor your father and your mother.
6. You shall not murder.
7. You shall not commit adultery.
8. You shall not steal.
9. You shall not bear false witness against your neighbor.
10. You shall not covet.

This is evolution, at work. Commandments such as "you shall not murder" and the like, have persisted throughout human history in all cultures simply because cultures which restrain our instinct to kill each other in a passionate rage and otherwise are the cultures that survived. Morality promotes social order and social order is the foundational human survival strategy, otherwise the ten commandments might include "Kill everyone except your descendants".

Imbuing our respective moral imperatives with a divine Imprimatur, causes us to feel righteous in our compliance and guilty (unless one is a sociopath) when we disobey. I understand this very well because as a child, I was raised a Fundamentalist Christian, I was taught that the only source of moral authority was the Christian tripartite God of the New Testament, that the bible was literally true and that to believe that Jesus was the son of God who died for our sins was a necessary and sufficient condition to gain one's place in Heaven and that failing to believe would damn one to an afterlife of eternal fiery torment in Hell. The concept that the world was ruled by natural law and that life on earth evolved naturally from nonliving chemicals was completely rejected.

The advantage for society in believing that injunctions against murder, stealing, false witness, adultery and other disruptive behavior are divine commandments makes them much more likely to be obeyed. Especially since we are all capable of breaking any of them in the right circumstances. I have often wondered if the author, traditionally said to be Moses, of the Ten Commandments really thought that they were written by God or was just being clever in claiming that they were. I have long had the suspicion that the latter was the case, because it is simply against the laws of physics and biology for them to be written on stone tablets by other than a human hand. Regardless, the Ten Commandments are believed to be the word of God by a large proportion of the human population, not just the Jews for whom they were written. This would not be the case if Moses had simply presented them as civil law that should be obeyed because they were good ideas.

Thomas Jefferson, who was not particularly religious and who edited out all the supernatural elements in the New Testament, recognized this. He wrote in our Declaration of Independence, that we are endowed by our Creator with certain unalienable rights, among which are Life, Liberty and the pursuit of Happiness. By doing this he made these rights into divine commandments even though they do not appear in any sacred text. I point this out to illustrate that religious doctrine is not written in stone, the Ten Commandments may be an exception. Believing that our rights are divinely granted has had a profound emotional effect on our citizens to the point that a great many have sacrificed their lives in their defense.

The importance of moral imperatives in controlling human behavior toward the Biosphere became apparent to me very early in my career. I assisted a wide range of individuals, groups, and units of government to make environmentally sound decisions about land use and other issues. I had hundreds if not thousands of conversations with people from all walks of life. It became clear to me that virtually everyone was guided more by their beliefs about the Biosphere than their scientific knowledge of it. Decisions were as much moral decisions as rational ones. In the United States our obsession with the rights of landowners is tinged with a deep moral certitude. For many this it is a God given right to do with one's land whatever one wishes. This, of course, can make it difficult to persuade a landowner to voluntarily refrain from activities that harm the Biosphere. While this comes from our basic instinct to defend a territory, an instinct we share with other species, our sentience expresses it with a clear reference to the Judeo/Christian and other religion's belief that we own the world and the creatures in it.

Even the beliefs of those who were not overtly religious, almost always had religious roots. Since I worked in a predominantly Christian country most of these beliefs were based interpretations of the Bible. But I am sure that an analogous spectrum of beliefs would be found in other religions. For example, countries that are predominantly Muslim make no secret of deriving civil law from interpretations of the Quran. Although western civilization has become more secular in recent times, we still refer to a loosely defined

Judeo/Christian tradition of morality. Likewise, China and other states that are or were nominally Communist, have not completely abandoned their respective religious heritages. We see this demonstrated in a resurgence of religion in countries that have overthrown communism, such as in the former Soviet Union.

Unfortunately, I found that what far too many nonprofessionals believed about the Biosphere was not only factually wrong but wrong in ways that were potentially harmful to it. This was not simply a case of being ignorant of the science involved, but of rejecting science outright because of deeply held beliefs. For example, the belief that economic benefits supersede environmental benefits is not only believed by individuals but written to U.S. law. Many people took the phrase "drain the swamp" literally as a good thing to do and were willing to fill a wetland for even trivial reasons. The problem with this is twofold, first economics is a poor way to value native ecosystems because it promotes the resource value of ecosystems rather than their value as functioning ecosystems. Second, economic benefits are calculated on a short-term basis, usually 50 to 100 years. Whereas environmental befits can be measured in geologic time. Such beliefs were so strongly held that my science-based arguments to the contrary fell on deaf ears.

What surprised me was that most people were unaware of the extent to which their decisions were driven by these deeply held beliefs. We accept the myth that we are rational creatures and that our actions are based on knowledge not belief. We are hindered by our genetic predisposition to believe that the world is under divine control. This is unlikely to change in the short time available to us to heal the Biosphere before we self-destruct. Belief of this depth cannot be changed by scientific arguments, no matter how well reasoned. If a person has been raised to believe that the world was created about 6,000 years ago and the first humans lived in the Garden of Eden, trying to change their minds is pointless and will likely anger them. Knowledge alone will never convince religious believers that life in the Biosphere is a natural process governed by biological evolution. Because of this, most the world's population will never accept a scientific world view, presented as such.

I believe that we must harness the power of religious belief to provide the motivation to heal the Biosphere. Because for most people religion forms the foundation of moral authority. Understanding this led me to the epiphany that neither science nor religion alone can save us from ourselves. Humanity will not commit itself to healing the Biosphere until a enough of those wielding political authority accept that it is a moral imperative to do so. For this to happen religious beliefs about the Biosphere must be informed by scientific knowledge of the Biosphere. In democratically governed states, this, at least in theory, means most voters must be on board, while in authoritarian states it means those in power. This approach has the advantage in that the faithful do not need to understand much of the science involved any more than patients understand intricacies of brain surgery, they need only believe that it is the right thing to do.

The key lies in expressing a scientific view of the Biosphere and its healing in the form of divine commandment. Only this will give the science we need to heal the Biosphere, the importance it needs in land use decisions. I call this "science informed religion." In which healing of the Biosphere must accepted as part of God's plan. This will require changes to beliefs or at least reinterpretations of beliefs to bring them more in line with a scientific understanding of the world. This will be much more persuasive to religious communities than scientific arguments.

I do not mean that we must convince those who believe in a religious understanding of the world to suddenly reject it for one that is science based. Theory of evolution and other theories conflict with too many deeply held beliefs to make that possible or even desirable. Science is a poor substitute for religion for it makes no moral judgements about its use. The building of the first atomic bomb could not have been done without science, but the moral decision to build it and ultimately use it was based on the belief that God was on our side. The Japanese believed that god was on their side in the form of a divine emperor. The Germans put "Gott Mit Uns" (God with us) on soldier's belt buckles.

What I am proposing is that we modify religious doctrine to permit the use of science in the healing process, even though the average nonscientist may not understand or even agree with the underlying scientific theories. This use of science by those whose beliefs are contradictory to science happens every day without controversy. An example is the use of global positioning systems (GPS) in automobiles. GPS systems depend on Einstein's theory of general relatively, which is rejected by certain fundamentalist Christians. Likewise, the theory of evolution is the foundation of medical research but is considered heresy by many who make use of modern medicine.

Existing beliefs are not enough for the job. First, they rest upon interpretations of ancient texts which give an incorrect picture of the Biosphere that cannot be used to develop science-based restoration plans. An example would be Pope Francis, who has apparently recognized that the world has a limited carrying capacity for humans has stated that Roman Catholics do not have to breed like rabbits just because artificial birth control is against church doctrine. While this is laudable, the fact remains that the church sanctioned methods of birth are limited and ineffective because they are based on the false premise that sex is ordained by God exclusively for reproduction and that sex for pleasure is contrary to natural law. But this is simply a misinterpretation of ancient science. It would be much more effective to accept that current science has provided us with effective birth control, thereby allowing sex to also be a healthy form of recreation. Apparently, many Catholics, at least in the U.S. understand this.

This will not be easy, most existing Religions do not accept that we now know much more about how the Biosphere operates than did the authors of ancient sacred texts and are generally resistant to changing doctrine in the face of scientific discoveries. Many religions go so far as to reject the very science on which healing the Biosphere must be based. Our genetic predisposition to believe the world is under divine control is unlikely to change in the short time available to us to heal the Biosphere before we self-destruct.

The good news is that it is not necessary to convince every believer any more than it is necessary to convince an entire flock of sheep to enter a pen. One need only convince the lead sheep. It is no coincidence that the congregants in a church are called the flock. The shepherds being priests, rabbi's, ministers, imams, shaman and other religious leaders. These figures are empowered by the belief of the flock. Humans need to be reassured by someone in authority that they are obeying divine commandments. Our task is to persuade those shepherds to reinterpret the doctrines of their respective religions in a way that makes healing the Biosphere a Divine commandment.

I am hopeful because throughout history, science and technology have been harnessed for religious ends. The development of calendars with which to time both secular and religious events being an example. The Gothic cathedrals of the middle ages were built because people who lived in squalor felt a moral imperative to glorify God rather than improve their living conditions. Their construction is astounding compared to the resources and technology at the time they were built. I would add the codicil that moving mountains, or any other large object is made easier by a scientific understanding of machines.

This is not a new idea, religions have long ascribed a divine cause to natural phenomena, or put another way, incorporated natural phenomena into their theology. We must expand those theologies to include the scientific principles necessary to carry out a science-based plan for healing the global ecosystem. This is especially true for the Theory of Evolution. To do that we must change what the religions of the world teach about the Biosphere. This is not deception, first because what science has discovered is the truth and can be verified as such, second since we do not and cannot know the truth of any religion, we cannot falsify its theology by changing it. The only thing we can know is what science discovers. We cannot know if a God or Gods exist, we can only believe.

Alternatively, it may be necessary to start a new religion based on the belief that what we discover through science is as close to divine revelation as we are likely to get. But establishing a new religion without referring to metaphysics or divine beings or any other figments of our vivid imagination, will take some thought. To begin with, a good religion needs a sense of mystery, and the real universe has more than enough mysteries for anyone willing to appreciate them. The weirdness of quantum mechanics, the relationship of mathematics to the physical world, how life began and the question of extra-terrestrial life, to name a few. None of these mysteries require a supernatural explanation, never-the-less they do inspire a certain sense of awe that so much could spring from the fundamental forces and particles of the universe.

I am not suggesting that scientists by themselves can change what Christianity or any other religion teaches about the Biosphere. What scientists can do, is to determine what beliefs need to be changed to make healing the Biosphere with a science-based approach a religious duty. The change itself must be brought about by the leaders of these religions. A healthy Biosphere must come to be understood as is a divinely endowed human right. It is the responsibility of religious leaders to incorporate this idea into their respective doctrines.

Restructuring religious beliefs in this way will encourage us to heal the Biosphere rather than justify our destroying it. I realize the difficulty of this task because of the fundamental difference between moral belief and scientific knowledge. Deriving the former from the latter is controversial if not heretical, especially if it means rethinking some religious doctrines in the light of what science has discovered about our universe. It will take forward thinking religious leaders to understand what is at stake and make this happen. It would be much simpler if we had an eleventh commandment to this effect, but alas we do not. The Ten Commandments and other religious texts were written to promote social cohesion through proper behavior, not to deal with the ecological problems we face on a global scale.

I am a scientist not a theologian and cannot say how this can be done. but I can offer some suggestions based on my own religious upbringing and experience in ecosystem restoration. Those of us that subscribe to the Judeo/Christian tradition might be led to believe that in addition to being endowed by our creator, as Jefferson proposed, with the inalienable rights of life, liberty, and the pursuit of happiness we are also endowed with the right to a healthy Biosphere. This would not contradict at least some interpretations of Genesis, which emphasize the stewardship of the earth and dispense with the harmful idea of dominating the earth.

Some religions might accept that our intellect and the science it has produced is a gift God has given to us to understand and care for the world he created. God allows us to use our intelligence to discover the truth about the Biosphere, which elevates the discoveries of science to` a form of divine revelation. This would imply that God created the world using the laws of nature and that biological evolution operating in the Biosphere is the mechanism for the creation of life and that all living things are its progeny. In that case, ancient Sacred texts would be understood as stories and moral lessons from a time before science. Many religions would argue that constraining creation to the laws of nature limits the power of God. My response is that it is not a limitation if God created the laws of nature to use in the creation. This will be the most difficult for prophetic religions that depend on specific events occurring in history, such as those described in Genesis. Those Christian denominations who accept Genesis as a book of moral lessons not a history text will have less difficulty in incorporating science into their theology.

These are but two approaches that might work in some cases. There are probably many others that would come to light following a serious investigation. I have no doubt that within Islam, Hinduism, Buddhism and other religions approaches can be found. The approach taken would depend on the peculiarities of individual religions. Fortunately, much scientific knowledge needed for ecosystem restoration does not conflict most any religious doctrine. Few would have religious objections to the fact that water runs downhill or that some soils are more fertile than others, one could think of thousands of examples like this. We need to build on this common ground.

What we must do is to work with theologians to understand exactly where the conflicts lay and see if there is a work around whereby believers can accept those parts of theories and facts which are necessary for restoration. Some parts of science will never be accepted and that may be fine if what is not accepted is not important to the task at hand. For example, scientific arguments will never convince bible literalists that that evolution is the rule of life or that the world is billions not thousands of years old. But we may not need to.

Creationists believe that dinosaurs coexisted with humans but are now extinct and scientists believe that their existence was separated by millions of years and are now extinct. The point is that both groups recognize that ecosystems change over time. That commonality could be very important because ecosystem restoration plans must take into consideration that what is restored is not static, rather what is restored is a set of processes which will change over time.

The fact that religious beliefs are interpretations of ancient text means that their doctrine depends on the interpreter. The idea that we should be stewards of the earth is appealing, however, the same text can be interpreted to mean that we should subdue the earth or that we need not worry about the earth because it is only our temporary home.

What is needed is a belief system which incorporates a moral approach to the restoration of native ecosystems but at the same time is not based on sentimentality about nature, or a metaphysics which is inconsistent with a scientific understanding of the functioning of the Biosphere. While it is acceptable to use a spiritual world as metaphor, it is counterproductive to act as though the universe is controlled by other than natural forces. This is true whether beliefs are ascribed as divine writ or as simply being in harmony with "nature". Many will argue that this is a fool's dream, and perhaps it is, but I think not. Humans accomplish great things in the name of belief. The great cathedrals of Europe were built by people living in mud huts because they believed that they were doing to be God's work.

A SECULAR MORAL IMPERATIVE

In the last chapter, I explained that moral imperatives rest upon an accepted moral authority. Most humans believe this to be some form of the divine. Such beliefs give moral imperatives a divine imprimatur, essentially converting them to divine commandments incumbent on all believers. Religions differ as to exactly why we should obey, some emphasize a love of God others the fear of punishment while others simply believe that obedience is our duty. Regardless of these theological differences the bottom line is that a believer must obey.

On the other hand, a minority of humans, myself included, reject such beliefs as wishful thinking. We understand that God, while seeming real to believers, exists only in the mind as a part of our subjective experience of life, and not in the external world where life occurs as an indifferent objective process. Like believers, we instinctively want there to be a moral authority to maintain social order, but if the universe is an indifferent process in which we are but one species among many what can possibly serve that function?

I believe the Biosphere itself can serve as our moral authority, but only if we accept it as science discovers it. Regardless of what ultimate reality is, and regardless of our individual wishes and sentiments, the Biosphere is our proximate reality and literally our creator. As its children must love and honor it even though it cannot love us back. The fact that the universe is indifferent to me as an individual does not prevent me from loving my family and contributing what I can to the betterment of society.

Within our subjective experience of life, we can infer that our purpose, at least metaphorically, is to play the evolution game to win, and in doing so, pass on our genome to our descendants, like racers in a relay race. Regardless of what ultimate reality is found to be the Biosphere is as it is because it is an emergent product of the physics of reality. It is governed by the rules of the evolution game and our individual and species survival depends on playing to win and winning means surviving to continue playing. Evolution has given us the privilege of being the test case for advanced sentience and sapience and we are obliged to use these adaptations to our best advantage. We can only succeed in that by restoring and maintaining the health of the Biosphere.

This is our purpose, not in a teleological way, there is certainly no evidence that evolution was predetermined to produce us, but in an emergent way in the sense that the processes that constitute biological evolution inevitably produce an intelligent species if given enough time. Exactly what our descendants will look like we cannot know but my hope is that we will evolve into a species that has more rational control over its activities. If we evolve in this way or if another intelligent and rational evolves from some other branch of the evolutionary tree, it would demonstrate the success of sentience and sapience as survival strategies.

SCIENCE AS A TOOL FOR HEALING THE BIOSPHERE

Considering life an objective process, it is clear to me that our species, like all others, is neither good nor evil, but simply a player in the evolution game. We are destroying the Biosphere because we are instinctively following the fundamental rule of the game, which is to use our adaptations to survive and pass on our genes. Our problem then is not one of good vs evil, rather it is that our evolutionary adaptations are a two-edged sword, the power of our intellect and manipulative skills has helped us survive in the short term but is endangering us in the long term.

We simply lack the lack the ability to control ourselves because our intelligence is not in control of our instincts. The only long-term solution is to evolve from being a species with intelligence to an intelligent species. An intelligent species being one whose intelligence controls instinct and not the other way around. But it is unlikely that such evolution will happen in time to save us from extinction. In the meantime, the Biosphere is in acute danger and we must act now. This is a scientific fact whether we want to believe it or not.

Many, if not most people, believe that our purpose for existence is part of a divine plan and that the origin and fate of the universe is in God's hands, I do not. As a scientist I believe that the universe is a collection of natural processes that came into being without divine intervention. As such we are free to decide for ourselves what our purpose should be, from the study of the universe itself. I chose to believe that our purpose is to demonstrate that sentience and sapience are successful evolutionary adaptions and that creatures such as ourselves possessing a high level of these attributes are not, *a priori*, doomed to bring about our own extinction. But to fulfill this purpose, we must first heal the Biosphere of the injuries we have inflicted on it using the science of ecosystem restoration.

I know firsthand the technical and social problems involved with changing our behavior toward the Biosphere. Never-the-less I am convinced, based on my experience doing ecosystem restoration on the ground, that we must fundamentally change how our species plays the evolution game. Because we have not evolved the capacity do this directly using our intelligence, we must do it indirectly using our intelligence to change how the world's religions view the Biosphere and then harness the power of this belief to change how we behave vis a vis the Biosphere.

I have seen how easy it is for the members of the public to be confused and alarmed when scientific knowledge conflicts with deeply held beliefs. Concerns about the freedom to use one's land as one sees fit, belief in the power of the free market, fear of government interference in people's lives often inhibited or even prevented the implementation of scientifically correct plans to address environmental problems.

In previous chapters I have tried to outline how these concerns might be overcome and our beliefs might be changed. But this chapter is not about belief, but rather about the scientific approach to healing the Biosphere that belief must motivate.

Healing the Biosphere will be the most difficult task in human history, it will take decades and the combined talents of our best scientists, engineers, theologians, politicians and other disciplines. Our goal must be to reduce human impacts on native ecosystems while at the same time restoring enough of them to achieve a balance between our needs and the ability of the Biosphere to meet those needs.

The scale of what must be done will be unprecedented and will necessitate sacrifices by both nations and individuals. It will require ecosystem restoration on a global scale recognizing that the world's native ecosystems, the forests, prairies, deserts and the rest, constitute the natural earth's natural infrastructure that provides the temperature, atmosphere and a myriad of other goods and services that make life possible on this planet.

The missions of agencies charged with protecting the environment will need to be changed as needed to reflect an ecosystem restoration approach rather than that is resource protection. I spent my career as a wildlife biologist for the Natural Resources Conservation Service, formerly the Soil Conservation Service. Not surprisingly we called our efforts to protect our environment "Resource Conservation". Our very name tied us to the idea that it is the components of ecosystems, e.g. soil, water, timber etc. that are separable resources to be conserved rather than the functioning native ecosystems which produce them. In fact, using the term ecosystem was frowned on. This overarching belief in the "wisdom of the marketplace" forced me to justify the preservation of native ecosystems with reasoned arguments while a land developer is free to invoke an emotional appeal to the value of economic expansion. This elevates the economic value of the land over the ecological values a wetland or other native ecosystem might perform, thereby favoring land development over ecosystem preservation.

Because we were an agency of the United States Department of Agriculture, we were charged with conserving agricultural land. To this end, we were originally named the Soil Erosion Service and later the Soil Conservation Service (SCS). SCS was formed during the dust bowl of the nineteen thirties. Our central mission then and now was to conserve the agricultural soil by protecting it from wind and water erosion, we were given this mission with the best intentions, but in practice it had, unexpected negative consequences.

While my agency was very successful in reducing soil erosion on agricultural land by the installation of certain farming practices such as strip cropping, these same practices came at an environmental cost. Strip cropping, certainly the most recognizable conservation practice, had a negative impact on ecosystem health because farmers enlarged their fields and removed hedgerows between fields to make room for longer continuous strips. Another conservation practice, no-till cropping which, as its name implies, reduces the need for plowing and otherwise disturbing the soil has the negative impact of requiring large amounts of herbicide and insecticide.

An example of the failure to recognize the value of intact naive ecosystems was the conversion of the tall grass prairie of the central United States to cropland in the late 19th century. The tall grass prairie ecosystem was systematically destroyed by the wholesale slaughter of the vast herds of native bison and the conversion of their habitat to cropland. This grassland ecosystem which existed for thousands of years could have provided a self-sustaining abundance of high-quality meat with little human intervention. Instead we plowed up the grassland and converted it to cropland in which corn is grown and fed to domestic cattle in feed lots. This cropland unlike the original prairie grassland is not self-sustaining, requiring large inputs of fossil fuel, fertilizer and pesticide to maintain it.

Admittedly there were some programs to restore portions of the tall grass prairie. These were primarily aimed at restoring pothole wetlands for the benefit of waterfowl, a valuable resource because of the money spent on hunting ducks and geese. These programs encouraged farmers to cease cultivation of uneconomic agricultural lands and had as much to do with subsidizing farmers as restoring viable ecosystems. Even those programs which were designed specifically to restore wetlands were very limited in scope in part because the prairies have been broken up by land ownership patterns and were hampered by anthropogenic changes at a regional and global scale. In some midwestern states, county wide drainage systems significantly change ground water levels and essentially prevent the restoration of the hydrology necessary for the restoration of wetland complexes, that is wetlands of varying "wetness" from ephemeral to seasonal to permanent. The disappearance the natural prairie seed bank, the introduction of invasive species of plants and animals, and the control of wildfires further limits the potential to restore native prairie ecosystems.

I remember a staff meeting several years ago. The subject came of streambank stabilization in agricultural areas. Basically, the problem is this. Farmers in New Hampshire and in the rest of the country like to farm along streams and rivers because the land is flat and easy to work. Typically, the best agricultural land is on flood plains. Farmers have an incentive to farm as close to the edge of a stream as they can because the more crops, the more income they have.

Rivers on a flat flood plain tend to meander forming ox bows. One side of a stream or river in this situation tends to erode while the other side builds up. Areas of building up and eroding alternate along the course of the river, causing the channel to meander back and forth across the flood plain. This a natural and dynamic process.

Streambank erosion and the consequential meandering of streams and rivers is a natural process in native ecosystems, Wetlands formed when oxbows are cut off, provide valuable wildlife habitat, storage for flood waters and many other beneficial functions. From the point of view of ecosystem health this is a good thing, part of what is called the erosion cycle in which mountains arise and subsequently erode. However, from the point of view of a farmer it is as a bad thing because of the loss of farmland.

The solution to this has been to straighten stream channels, particularly smaller streams and rivers, and armor the banks with rocks to prevent meandering. This of course makes life easier for the farmer but seriously disturbs the ecosystem process of stream meander. Our agency spent much time trying different approaches to stop this erosion by what was called streambank stabilization. Among the solutions that were tried were lining the stream bank with rip-rap, i.e.

large rocks and planting varieties of vegetation such as willows bred for their ability to grow rapidly and hold soil with their roots. The unexpected consequences of streambank stabilization are several. The fast-growing willows can become an invasive species, crowding out native stream bank vegetation. Armoring streambanks changes the hydraulics of a stream and stops the natural meandering of the stream, as well as simply passing erosive forces downstream.

During the meeting, the discussion turned to the subject of streambank stabilization. I put in my two cents that stream channel erosion was a natural part of the dynamics of flood plain ecosystems and while our agency should try and reduce excessive erosion along stream channels we should not try and stop the natural erosional process from occurring. I also made that point that by and large the erosional process is necessary for other ecosystem to function properly, for example river delta's need sediment.

Another staff member, who happened to be a specialist in soils, said that no, I was wrong in not wanting to stop meandering on these streams in agricultural because *soil (in this case agricultural soils) were the resource we were mandated to conserve.* Here you have it in a nut shell. The emphasis on soil in the agricultural sense as a resource to protect could lead if carried far enough to the disruption of not only flood plain ecosystems but other valuable ecosystems downstream.

My agency also produced county by county soils maps of the United States as an aid to farmers. These maps describe the characteristics of individual soils that are important for farming such as its depth, moisture holding capacity and stoniness. The intent was to conserve agricultural land by using soils information as a tool in farming. While at first glance this seems like a good idea, the unexpected consequence was that these maps are just as often used by land developers to identify sites for building houses and malls, based on soils characteristics.

But there is another way to understand this text. Considering that there is a lack of hard evidence that Genesis is God's word written down by Moses, the text makes more sense if you turn it around and understand that the bible is not God's word written by down by Moses but a human document, as are all other so-called sacred texts. These sacred texts are not the cause of our destructive behavior, which is instinctive, rather they are our cover story to justify what we do. Later books added to the bible by Christians compound the problem by making it clear that the world is only a temporary stop on the way to eternity, a sentiment which does not encourage our taking good care of the Biosphere.

Believing this we simply don't take the long-term impacts of human activity on the Biosphere seriously and only give lip service at best to any thought of restraining human activity to a sustainable level. We reflect this disregard for the future at a governmental level where we typically have a planning horizon of 100 years or less for public works projects. This is a millisecond on a geologic or evolutionary time scale. We think nothing of destroying important ecosystems for a hundred years of flood protection, which may fail in an even shorter period.

The exploitive approach to ecosystems is also evident in the training of many professionals. It is a given among foresters, that forests are to be managed for timber, regardless of the impacts on wildlife or native plants. Wildlife biologists have traditionally managed forests for specific game species again regardless of the impact to the ecosystem. Agronomists and plant scientists think nothing of breeding or importing highly aggressive plants for some purpose and only later discovering that these plants have become invasive and are crowding out native species. Attitudes such as this are embedded throughout the professional world to the degree that they are assumed to be part of the natural order of things; simply common sense and not to be questioned.

The attitudes of engineers, economists, wildlife managers and others I encountered were often very narrow and limited to their respective disciplines or personal beliefs. A meeting of resource professionals often reminded me of the ancient fable of blind men describing an elephant. Each was very good at describing a limb, trunk or tail by none had any idea of the whole animal. This type of misunderstandings among professionals could be chalked up to differences in how their respective disciplines view ecosystems. I remember a colleague who believed that we should use up the world's resources as fast as possible because doing so will hasten the end of the world and the second coming of Jesus. This belief, which he sincerely held, is about as far from the idea of stewardship as one can get. This of course was a bit disconcerting to me as we were both employed by an agency whose purpose was to promote the conservation of natural resources and his specific job title was Resource Conservationist.

Ecosystem restoration is in many ways, analogous to the practice of medicine. The Restorationist, like the physician is obligated to first do no harm. It requires both formal training and field experience because it is both an art and a science. A medical practitioner typically takes a medical history, asks about symptoms, examines the patient and performs indicated tests.

Then, based on his or her training and clinical experience, makes a diagnosis and prescribes a treatment. An ecosystem Restorationist follows a similar approach only the patient is a native ecosystem and the disease is the reduction in ecosystem structure and function caused it by human activity. The Restorationist must research an ecosystem's history, examine the ecosystem in the field looking for the symptoms of human activity, such as the presence of invasive plant, fills and excavations, changes in hydrology due to drainage and a myriad of other impacts. The Restorationist must also understand the ecological history of the planning area. How did the ecosystems we see today come to be? How closely do they match the expected native ecosystems of the area? What forces have shaped them? How have humans affected them? What will the future condition of the ecosystems be, with and without human intervention? What is the potential for restoration?

The Restorationist must also perform any necessary tests, such as screening for toxic waste, inventorying plants and animal populations and so forth to understand its current state of health. Then, like a medical practitioner, make a diagnosis and prescribe a treatment in the form of a restoration plan. As in medicine, it is often necessary to consult with specialists in other fields, including hydrologists to assess changes in tidal flow and the impacts of flood control and engineers to make the necessary physical changes such as removal of fill and rerouting of utilities.

Just as in medical practice the stakeholders in any restoration plan must have enough faith in the practitioner to consent to the treatment. This is especially true if the treatment involves pain and risk. In the case of Biospheric restoration the prescribed treatment will involve both. Healing the Biosphere must also be an ongoing process, accompanied by a gradual reduction in global population, through family planning, and the gradual relocation of human populations away from critical ecosystems.

Throughout the United States and Canada local ecosystems exhibit similar anthropogenic problems, including reduction in biodiversity, desertification, hydraulic and hydrologic changes, soil degradation, habitat loss, local extinction, accelerated cropland and stream bank erosion, point and non-point pollution, invasive species and the like, the list is long. While the types of problems, in different locations, the specific soils, climate, slope and plant community vary widely from deserts to marine ecosystems. Restoration plans will have to be tailored to local conditions.

Unfortunately, we do not know is how much natural infrastructure is enough and if we continue our current trajectory, we may not find out until it is too late. We can only hope we have not already passed some unknown threshold beyond which the Biosphere is not repairable. The details of what will be necessary is beyond the scope of this book. But will include such things as rethinking how and where we live, build cities and farm. We will also have to address known problems by reversing desertification, reforestation, relocation of populations, redrawing of political boundaries and extracting plastic waste from the ocean to name a few.

In practice, restoring the health of the Biosphere will be a process of mitigating the stressors we have placed on it. We do this by restoring or mimicking as much of the structure and function of native ecosystems as necessary to return the Biosphere to a sustainable condition. While we can rely on natural processes to help with this, our active intervention will be required. We cannot simply walk away from a degraded ecosystem and expect it to heal itself. In most cases, the more "natural" we want an ecosystem the more management it will require. This may seem counter intuitive, but the sad fact is that we have so disrupted the natural processes of most native ecosystems that they can no longer be restored without human intervention. The filling of wetlands is but one example. Others include the reduction or blockage of tidal flow in saltmarshes and estuaries, over harvesting of timber and of course building cities and other human infrastructure. Changes in soils caused by farming and other activities such as strip mining, construction and deforestation. The list goes on and on, but the bottom line is that in many such cases we must physically change a given area to restore or mimic lost natural processes and maintain the restoration practices we have applied to it.

Relieving anthropogenic stressors on at a global scale will require changes in behavior of people living far away from the ecosystem being restored, because all the earth's ecosystems are interconnected in ways we do not fully understand. Events on one side of the globe can affect ecosystems on the other side. The Gulf Stream originates in the Gulf of Mexico but the warmth it brings to the northern regions provides the British Isles with a moderate climate. The restoration of an estuary or embayment may require changes in agricultural practices and municipal wastewater disposal methods in its watershed far upriver. The support tacit or otherwise of people whose actions stress a given ecosystem are essential, even if these people do not live in or otherwise derive any direct benefit from the ecosystem being restored.

One of the major problems we face is that of positive feedback loops, which are situations in which the solution to a given environmental problem produces a greater environmental problem that it solved. Virtually all environmental problems are therefore result from the solution to a former environmental problem. Fire and agriculture being two prime examples. Both initially solved the important problems facing early humans, cold weather, the difficulty of finding food and the necessity of having to eat what food was found raw. Unfortunately, the use of fire and agriculture produced positive feedback loops in that both resulted in increases in population which demanded that forests be cut down for fuel and prairies be plowed up for agriculture, which in turn resulted in even larger populations and greater demands on the Biosphere.

Even in antiquity cities such as Rome required the importation of grain from Egypt as well as the conversion of local native ecosystems to agriculture. In modern times, large scale farming which depends on the use of fire to clear land and make internal combustion engines run has had negative impact on ecosystems worldwide. In the United States farming has virtually destroyed the tall grass prairie of the Midwest and other largescale ecosystems such as the hardwood forests of the Mississippi River Valley. The conversion of prairie into cropland has eliminated the natural seed bank over large areas as well as virtually eliminating the animal communities, such as prairie dog towns and Bison herds that maintained the prairie. Farming has additional negative impacts including but certainly not limited to reducing soil carbon, depleting ground water and introducing invasive plants. In addition, large scale agricultural drainage, irrigation and flood control projects have significantly altered the hydrology, further reducing the chances of restoring native ecosystems on agricultural land.

The result is that there are simply too many humans and too few fully functioning native ecosystems. This should be obvious to anyone who reflects how much of the earth's natural infrastructure we have eliminated to feed and house us. We must set human population goals to a level that will allow the restoration of enough native ecosystems to maintain the earth as sustainable human habitat. Unfortunately, this problem is so bound up in our current social, cultural, religious and economic systems that at first glance it appears intractable. Nevertheless, overpopulation must be addressed by vigorous worldwide birth-control measures.

The difficulty of ecosystem restoration projects is a function of scale, the larger the project the more difficult it is. This is obvious; restoring a small wetland is easier than restoring the Chesapeake Bay. I recognized early in my career that most government programs that I was involved in that were meant to help the environment seldom looked at the "big picture", that is their impact on a regional or global scale. State and Federal natural resource agencies are simply not mandated or equipped to deal with large scale issues such as sea level rise and climate change. National boundaries also reduce the possibility of cooperation on projects at a global scale. There are no international agencies empowered to work on restoration projects across borders.

We must also take a long view of maintaining the health of the Biosphere. Because evolution works in the short term and we tend to value those things we consciously or unconsciously believe will aid us to that end. This drive forces us to most often act in the short term either as individuals and as a species. This is reflected in the time frame for planning civil works such as road, bridges and flood control measures. Rarely do we plan beyond 50 to 100 years for such projects and we don't even try and access the effect of a road or dam or any structure or land use change 500 years or longer in the future. Five hundred years may seem a long time, but the fact is that we will need the oxygen produced by intact forest ecosystems centuries into the future, but we convert those same forests into cropland simply because of a rise in the price of a commodity crop such as soybeans, only later to find that the soil is not suited for cropland. This was the fate of the bottom land hardwood forest of the Mississippi river alluvial valley.

My own restoration work on salt marshes in New England, was local in scope may be wiped out in the future by global sea level rise. The major current problem was blockage of tidal flow by roads. We could mitigate this by installing larger road culverts. This was highly effective; however, it did not address the problem of currently rising sea levels, which will cause and have caused in the past the inland migration of salt marshes. In New Hampshire, for example, saltmarshes have migrated inland for the last 5,000 years of so as evidenced by remnants of freshwater bogs beneath them. This of course did not pose a problem in precolonial times because the indigenous population did not depend on permanent infrastructure, as sea level rose, they simply moved inland. We, on the other hand, do build permanent infrastructure in abundance. This poses a problem for the future of salt marshes. Where will they migrate to if houses and other structures are in the way. This means that to really restore these marshes over the long term we must prepare now for this migration by removing structures, raising bridges and preventing land developing in low lying areas along the coast that have the potential to develop in to salt marshes in the future.

Originally, my agency was named the Soil Erosion Service, later changed to the Soil Conservation Service. Our mission was to work with farmers to install what were called Conservation Practices. Conservation Practices were farming techniques designed to reduce soil erosion by wind and water. During the dust bowl, the major problem was wind erosion. The name dust bowl alludes to the great clouds of dust that swept the corn belt of the United States at that time. The dust was eroded soil from the millions of acres of otherwise fertile cropland that was devoid of vegetation due to crop failure. Thousands of farmers and others were forced off their land because bank loans could not be paid. The drought had killed what little plant cover remained on cropland allowing the wind to blow away the top soil. This drought was particularly hard on the affected area because it lay west of the 100th meridian and even under average conditions, the evapotranspiration rate exceeds precipitation.

The dust bowl is still referred to as a natural disaster, but it was not. It was the result of the way in which humans understand and treat native ecosystems throughout history and the world. The native ecosystem that greeted the early settlers of European ancestry who arrived in the early 1800's was one of the most magnificent prairie ecosystems the recent earth has ever seen. The dust bowl and response to it is a good example of how our instincts sanctioned and promoted by religious beliefs can have a devastating and completely unnecessary impact on a native ecosystem that could have produced a virtually unlimited quantity of high quality meat for the foreseeable future with very little human input. Instead, we plowed up the prairie and replaced it with monoculture row crops having little genetic diversity and soil holding capacity.

Before the introduction of agriculture, what is now the corn belt was a prairie ecosystem suited to the relatively dry conditions. The invention of the steel bottom plow by had made the conversion to agriculture possible. Following the introduction of farming, made possible by the invention of the steel bottom plow, the thick layer of grass sod was broken up and row crops such as corn were planted. The native bison herds, which had migrated into the region during the ice age, were also virtually exterminated. Domestic livestock was also introduced as part of the conversion to agriculture. This was seen a good thing when it happened in the late 1800's but in retrospect it would have been better to have preserved the prairies as a viable ecosystem as a much more sustainable source of food.

Certain cities will have to be abandoned or at least significantly reduced in size. This is especially true of cities such as Los Vegas and New Orleans, which, because of their unsustainable location, should never have been built in the first place. But it is often possible to restore certain portions of the native ecosystem such as stream corridors and wetlands. Salt marshes are maintained by tidal flow and new tidal creeks may need to be excavated to replace those lost through road construction and other development. Pine Barrens and other subclimax communities need periodic fires. Since humans generally suppress natural wildfires, prescribed fires will often be needed. Alluvial forests and woodlands such as those that were once common in the lower Mississippi river valley need periodic flooding which must be restored. In short, our management of natural ecosystems is aimed at restoring, or at least mimicking, those natural conditions under which the system once flourished. Doing so will also have the beneficial effect of reducing the impacts of human activity on surrounding native ecosystems.

it is often possible to restore certain portions of the native ecosystem such as stream corridors and wetlands. Salt marshes are maintained by tidal flow and new tidal creeks may need to be excavated to replace those lost through road construction and other development. Pine Barrens and other subclimax communities need periodic fires. Since humans generally suppress natural wildfires, prescribed fires will often be needed. Alluvial forests and woodlands such as those that were once common in the lower Mississippi river valley need periodic flooding which must be restored. In short, our management of natural ecosystems is aimed at restoring, or at least mimicking, those natural conditions under which the system once flourished. Doing so will also have the beneficial effect of reducing the impacts of human activity on surrounding native ecosystems.

We define a potential native ecosystem as the expected ecosystem at a given location following restoration of the physical, chemical, and biological conditions existing prior to its being significantly altered by humans. For example, if a wooded swamp has been cleared, filled and built upon, its potential native ecosystem may be wooded swamp. In other words, it would be possible, though costly, to restore it, to pre-colonial conditions. This means that every square inch of New Hampshire inherently has a native ecosystem associated with it because of its physical characteristics such as location, climate, proximity to the ocean, etc. Any given area would exhibit its characteristic native ecosystem in the absence or removal of human influence.

By describing these parameters at any point on the ground, we can predict the ecosystem, which would exist and probably did exist under natural conditions. Imagine, for example, a landform that is an isolated glacial depression that has a wet (hydric) moisture regime, and substrate of muck and peat. This physical environment would result in the development, under natural conditions, of a characteristic plant and animal community that we commonly call a kettle-hole bog. Very few ecosystems can be restored to a pristine condition. Most ecosystems have undergone such significant changes that they can never be put back exactly as they were. Extinction, climate changes, sea level fluctuations, and a myriad of other reasons make returning most native ecosystems to a historically pristine state impossible. Wolves, American chestnut, elm, passenger pigeons and a host of other plants and animals are gone, or nearly so. Our restored native ecosystems must get along without them.

In extreme cases, native ecosystems cannot be fully restored because of irreversible changes to the landscape, for example seasonally flooded forests, such as occurred on the Mississippi alluvial valley cannot be fully restored on agricultural land because of large scale flood control along the Mississippi River. In these cases where human infrastructure cannot be removed, we must attempt to limit the impact to surrounding ecosystems and if possible restore those pieces of natural infrastructure which run through it. For example, restoring channelized streams and daylighting steams that have been put in culverts. Creating wildlife corridors and greenways within the city. Reducing runoff by storm water management, especially using porous pavement and detention basins. Reducing a city's consumption of fossil fuel and potable water. Maintaining green belts around cities and zoning to limit their spreading across the landscape. These are but a few of the possible measures that can be taken to lessen the impact of cities. In some cases, entire cities need to be abandoned as unsustainable, desert cities such as Los Vegas being an example.

We cannot simply walk away from severely damaged ecosystems and expect them to heal themselves. In most cases we will need to apply carefully designed measures aimed at restoring the missing ecosystem components and processes. Some of these measures will involve large scale engineering projects to reshape and restore landscapes. But we must use caution and only resort to engineered systems to replace parts of an ecosystem as a last resort. The reason for caution is that engineered systems and natural systems differ in fundamental ways and are seldom interchangeable, especially on a global scale. It would be impossible to artificially freeze the polar ice cap, better to reduce greenhouse gas and let it freeze naturally.

In practice, we must be satisfied with restored native ecosystems that are as natural as practical but not necessarily pristine. It also means that we should reduce human stressors on highly disturbed native ecosystems that underlie urban areas and cropland as much as feasible. In practice, it means that we attempt to create conditions that allow a target ecosystem to function and evolve with as little human help or interference as possible. It is restoring and maintaining the physical, chemical, and biological conditions necessary to allow natural ecosystems to function and evolve over time. Simply, it is reducing and reversing unnecessary human impacts on native systems. This is the heart of ecosystem restoration obviously this is simple to state but extremely difficult to carry out just considering the mechanics of restoration, much less the social impacts.

Finally, no matter how we proceed, we must keep in mind that there is only one Biosphere. We have nothing to compare it too or experiment on, and most importantly no room for error. In any case, we cannot replace its function, much less its natural beauty, with engineered systems on a global scale. Neither can humanity retreat underground or move to Mars. We are essentially tinkering with the only life support system we have, but, I would point out that we are already tinkering with the Biosphere, and not in a good way. We must not be deterred by what now seem unsolvable problems. The bottom line is that we must exercise caution and be bold at the same time because we are in a race against time.

EXPECT A RELIGIOUS BACKLASH

Why do so many people cling to the preposterous and patently incorrect religious beliefs about the origin and operation of the universe? The simple answer is that believing is easier than knowing. Even a child can believe Genesis chapter 1, but only an adult can understand the calculus of quantum mechanics, and that only after arduous study.

Religious leaders in my opinion use the gullibility of the ignorant for their own aggrandizement. Even so, I do not propose that humanity abandon existing religions for they give comfort to many and do provide a wide array of charitable activities. It would also not be a good idea from the public safety point of view. Religions help us keep our inborn viciousness at bay, but as I have tried to point out existing religions are at best ineffective in protecting the health of the Biosphere and at the worst very harmful to it.

I apologize in advance to those sincere believers whom this book will offend. Unfortunately, such offence is unavoidable because of the fundamental conflict between science and religion. Those who believe otherwise are misinformed about one or the other. Science deals with reality as we find it and Religion does not. The deities of the major religions cannot be proven to exist in this reality and therefore are, for all intents and purposes, imaginary.

I admit that it is disingenuous for me to propose that religions incorporate scientific truth into their doctrines. But I see no alternative. Most of the world's population believes in some sort of creation myth and are likely to continue to do so for the foreseeable future. Reasoned arguments will not convince them to accept a scientific view of the world. The science needed to heal the world must seem divinely revealed not discovered empirically. If this is a lie it is no worse than telling ourselves for eons that we are special creatures and can do as we like with the world, and certainly better for the Biosphere.

Nevertheless, religious fundamentalists of all faiths will likely be inflamed by what I suggest, given the power of our primitive drives to control our actions and our unlimited capacity for self-delusion. Those who attempt to heal the Biosphere must be prepared for the anger and possibly personal danger from religious zealots. But healing the Biosphere is critical to our long-term survival, therefore religious doctrine must be made and taught by leaders that it is God's will that such healing be done.

Appendix 1: Myths that harm the health of the Biosphere

Questioning our Sacred Myths

Socrates is the first recorded questioner of belief. His statement that "an unexamined life is not worth living" is now a cliché. But it does capture what I think is the essence of what it is to be sentient and by extension to be human. The idea that one could examine his or her life through the facility of reason was counter to every tenant of society up to that time. This is especially true of his idea that one could discover the proper way to live through a minute examination of how one lives

It was a pivotal point for humanity, or at least what we loosely call western civilization. I am sure that similar thoughts occurred to individuals throughout the world at the beginning of other advanced civilizations. I believe that humanity steps out of the constraints of instinct at the point where introspection begins.

Arguments of social justice aside, we seem to be a highly successful species. Even in the face of war, disease and famine our population is growing. We have spread out over the entire globe even into habitats that would be unavailable to us without technology. But this success is short term on a geologic scale. Humans really got going in a serious way after the last ice age which ended about 10,000 years ago. The question is whether this success with all its negative effects on the Biosphere can continue very far into the future or will our ecological IOU's come due at some point? The person who understands the science of ecosystem restoration and believes that we have a moral imperative to restore ecosystem health is equipped to influence our collective future. We will succeed only if we as humans, or at least enough of us to make a political difference, accept as a moral imperative ecosystem restoration and the other steps necessary to restore the biosphere to health.

It seems to me that if our concept of humanness is to have any meaning, we must be successful in the game of evolution. Humanness must, like a Tiger's "Tigerness", be an asset for survival, not a liability. This does not mean that we must survive as a species forever. Evolution will have its way and we cannot predict what sorts of creatures our genetic line will give rise to, but we will produce no descendants if we eliminate ourselves from the earth's gene pool in an instant of geologic time.

Sacred Texts Contain Revealed Truth about the World

Revealed truth is that found in sacred texts and myths. It is the truth because these texts are believed to be inspired by God or in some cases the actual words of God. The problem is that the truths contained in these texts about the origin and nature of the world are wrong. We know from investigating the world that it is an object in space-time governed by the laws of physics. Whether or not the laws of physics were written by a divine being is beside the point. It is the scientific and not the metaphysical truth we must understand and act on if we have any hope of surviving as a species.

Despite this, myths about the world are as important to motivating us to either protect or damage the Biosphere as any scientific knowledge we have about its functioning. Sadly, most people have a very poor understanding of how the Biosphere operates and its importance for life. Creationists and others who harbor unscientific beliefs about the Biosphere simply ignore scientific facts and act based on myth alone.

God or Other Divine Entities Created the World.

The world of the Christian Bible provides an example of this myth. In Genesis, God makes the world in seven days, makes all the creatures and the first humans. He then has Adam name the animals and gives him dominion over them. So how does this myth lead to neglect if not outright antipathy toward the Biosphere? Genesis contains two contradictory creation stories and was probably never meant to be taken literally. Neither story is not based on the observable facts of the world. Both are obviously the creation an inhabitant of the ancient world. At that time little was known about the physics, chemistry, biology and ecology of the world. The universe described in Genesis is shaped by the fears and superstitions of the time. That world, which bible literalists, cling to against all scientific evidence, is simply not the world we inhabit. Therefore, it is almost impossible for a believer who takes Genesis or any other fantastical creation myth literally to understand much less accept the processes which created of the world and the evolution of its organisms.

The important point is that the world of sacred myths bears little resemblance to the world as it exists. Rather it is child's play dough world in which nothing is really connected to anything else. Plants and animals are just created willy-nilly with no understanding of their evolutionary or ecological relationship. It is obviously a world created in the mind of an ancient author who had no knowledge of the facts of the world. God is a big kid sitting on the floor, shaping the firmament out of celestial clay. The result is a world that can be controlled by God like a child who manipulates clay figures.

The natural laws of the world of Genesis are the opposite of the actual laws which govern the Universe. The time scale off by billions of years. The concept of time itself is contrary to the facts of relativity. The size of the universe is greatly underestimated and so forth. The biology and ecology of the world is so utterly wrong that it is hard to know where to begin to debunk it. Whole books have been written on that subject. The idea that God can arbitrarily interfere with the physics and biology of the world without cascading ramifications is absurd.

The Rules of Evolution and Ecology don't apply to Humans

Once humans began farming, the environmental impacts of society increased exponentially. The most obvious effect of agriculture are on native plant communities. Tree cutting, grazing, draining, burning, plowing, tilling, planting, weeding, irrigation, flooding, plant selection and hybridization are common agricultural practices all of which have profound negative affect on native plant communities. The adoption of agriculture has allowed human populations to expand way beyond the levels that can be supported by hunter/gather societies. Freeing many people from the daily necessity of gathering or hunting food has allowed the development of complex societies and advanced technologies. The result of agriculture has been the modern world.

All of this makes it seem that humans have escaped the rules of evolution. While that may be true in the short term our escape depends on an agriculture that is not sustainable. Aside from the fact that agriculture depends on fossil fuel to till the land and harvest the crop, agriculture causes major changes in the characteristics of the soil itself. Nutrients and organic matter are often depleted, and poor farming practices causes the erosion of agricultural soils. Much of the agriculture of the United States depends on ancient ground water which is used up faster than it can be replenished resulting in the lowering of ground water levels. If this continues much land which is now farmed may be lost in the future.

But, you say, "Humans are a part of nature. How can you say that ecosystems dominated by humans are not "natural"? The answer is that humans are both part of nature in the sense of being part of the Biosphere, but they are also apart from nature in some very profound ways.

Simple people, uneducated people, need simple explanations of how the universe and living things came to be. I do not imply that there is anything wrong with this or that the uneducated lack innate intelligence to become educated. It is simply that millions of humans live the barest of existence, without the luxury of the educational opportunities we in the developed world take for granted. We who live lives in comparative ease are in no position to judge those who do not possess our understanding of the natural world.

Our common enemy is ignorance not the ignorant. If a person struggling to survive helps to harvest tropical forests to provide subsistence for his family, what is he guilty of? Is he anymore guilty of destroying the planet that we who drive our multi-ton gas gulping automobiles? What he does he does out of necessity, we often do out of self-indulgence. Many of those who live in developed nations, lack the motivation or cultural inclination to pursue an education in science. To explain to them the details of scientific point of view and the science-based arguments for the protection of native ecosystems is not realistic. This places a responsibility on those educated in a scientific understanding of the universe and the evolution of life to translate or at least aid in the translation of this understanding into forms, which can be understood by the uneducated

Engineering can Replace Ecosystems

This myth is particularly damaging, but it is easy to understand how people came to believe it. At a small scale we can do a lot with engineering to replace natural systems. Agriculture replaces gathering wild plants. Dams and Dikes can control flooding to some degree. We can build buildings and make clothes that allow us to live in inhospitable habitats. At the scale of the Biosphere, engineering has done little more than disrupt those global systems which are so vital.

The composition of the atmosphere is a good example. Our atmosphere is not happenstance. The gases in the atmosphere are continually being cycled through living organisms. The earth's atmosphere is about 20 percent nitrogen and 80 percent oxygen with traces of other gases such as carbon dioxide. This oxygen rich atmosphere which allows for the evolution of complex organisms would not exist were it not for photosynthesis. As animals convert food into energy, they produce carbon dioxide as a waste product. Green plants use the energy of sunlight to produce glucose and oxygen from this carbon dioxide. This replenishment of oxygen is necessary because it is a very reactive gas and would otherwise be tied up as iron and other oxides.

Photosynthesis also takes carbon dioxide out of the atmosphere. The nitrogen in the atmosphere is a product of the breakdown of proteins by living organisms. This nitrogen is recycled back into protein by nitrogen fixing bacteria in the roots of legumes such as beans. I would be difficult if not impossible to replace these natural cycles with engineered processes at a global scale.

We have Dominion over the World

The main problem with the Myth of Genesis and other ancient creation myths is the idea that the world was created for the benefit of humans or the related idea that God has given our species dominion over the rest of creation. Meaning that the world is ours to do with as we please. This is balanced in some Christian sects by the idea of stewardship of the earth. Meaning that we are taking care of the earth for someone else namely God. Given the history of environmental exploitation that has occurred in predominantly Christian nations it seems to me that bible-based stewardship has been more talked about than done.

The reality is that the universe our creator. We arose from stardust, not at the whim of a lonely old God who needed little animated mud people to perform like puppets for his amusement, but as an emergent property of a magnificent and deeply fascinating process. That we can now view this process and see it with reverence and admiration is one of its most amazing features. It is literally something from nothing or almost nothing.

Be Fruitful and Multiply

The belief that that the earth has an unlimited carrying capacity for our species is simply wrong. Our high population, particularly in areas such as deserts places great stress on freshwater resources. In the United States the City of Los Vegas is a prime example. It is maintained by water from Lake Mead, an impoundment on the Colorado River. In addition to the water consumed in this desert city, Lake Mead itself lost over 700,000-acre feet of water per year to evaporation between 1953 and 1990. The recent drought in the western United States has only made the situation worse.

There is also the limit to the amount of cropland available in the world. Much of the world's cropland is farmed in ways that are not stainable. In the American Midwest the reliance on ground water for irrigation cannot continue indefinitely. The Ogallala Aquifer on which more than one quarter of domestic irrigated cropland depends is being used up at a rate greater than it is being replenished.

The World is a Temporary Stop on the Way to Eternity

I believe that all humans regardless of ethnicity or religion want to believe that they possess a soul which is independent of our physical bodies and lives on in some form after our death. The idea that there is a part of us which is separate from our physical bodies has long been a central theme in both religion and philosophy. Certainly, the idea of an afterlife is comforting. Personally, I would like to believe it, particularly as I age and find my physical body failing a bit at a time. I sincerely would like to believe that I will somehow go on forever after I die a physical death. Given what we know about life and the universe in general it seems to me that a belief in an afterlife is really wishful thinking. But I cannot be certain one way or the other.

There are stories in the New Christian Testament of persons including Jesus being raised from the dead, but I personally find these unreliable, because at the time they supposedly occurred, we had much less of a scientific concept of life and death. People of that time lived in a world of fear and superstition and believed a whole host of things we know are simply not true. There is also the so-called near-death experience of people who have technically died for short periods of time. I do not think that these provide any information for or against life after death. Their heart may have stopped but they are not completely dead.

To be honest, I can think of many ways that I could be convinced that the soul lives on after death, but all I am offered are old stories and evidence that is explained in more prosaic ways. Consider, for example the issue of remembering past lives. Some traditions maintain that we go through many reincarnations; a sort of a spiritual house flipping. Why is it that we have no memories of these experiences, except under what is claimed to be hypnosis? When I was growing up there was a fashion believe in reincarnation in this country during the 1950's. Reincarnation is not part of Christian tradition, which says you live once and end up in heaven or hell, or in the case of Roman Catholics you have two other possible outcomes, limbo, which is being rethought and purgatory which is not.

The bottom line for me is that while none apart from those already dead can prove or disprove the existence of souls and an afterlife. As far as science is concerned there is no need postulate the existence of either as the answer to any biological riddle. All biological questions from the origin of life to consciousness can be answered within science without recourse to supernatural explanations.

The end of the World is at Hand

In some sense the end of the world is always at hand. Random cosmic events such as a large asteroid hitting the earth could indeed bring about the end of the world. This is different than the unfounded belief that God is going to soon end the world for reasons of his own. In the Christian tradition the end of the world has been predicted since the beginning of that religion. Those who are skeptical of the validity of prophesy are still waiting.

Unfortunately, if a person believes that the apocalypse is at hand, he or she has little incentive to protect the Biosphere. If on the other hand one believes that we are a species which can make important choices about the future of the planet, and that science can help us make choices which will increase the likelihood of human survival long-term, then he or she will likely act to restore the health of the Biosphere.

The Personification of the Biosphere (E.g. Mother Nature)

The personification of nature has a long history. Perhaps back to the first humans. We do not have clear archeological evidence of the first attempts to identify nature with a god but some of the earliest religious figures are of what appears to be a Mother goddess. We do not know exactly what these figures represent but it is reasonable to assume that they were a way of giving nature a form which could be worshiped. Later we see numerous cults arise around gods and goddesses representing all aspects of nature from individual animals to the monotheistic creator god of the Abrahamic religions.

The problem with beliefs which personify nature is that they are simply wrong. And this wrongness can lead us to treat the Biosphere in harmful ways. Nature cannot be personified because "Nature" does not exist as a thing to be personified. There is only the process of evolution which shapes ecosystems and life itself. Biological evolution cannot be personified because it is an indifferent physical/biological process.

By the concept of Mother Nature, we are giving a personality to something (nature) that does not have a personality. To have a personality you must be a sentient being. We have personalities because of our having advanced nervous systems. There is no evidence that the Biosphere is an organism much less a sentient being. There are unproven arguments to the contrary, the Gaia hypothesis being one. Even this hypothesis does not credit the earth with being a sentient being.

Accepting That Nature is an indifferent process is difficult. Evidence the number of individuals who subscribe to belief systems in which God or Nature itself cares about the fate of humans to the point of intervening in the universe through miracles. We want nature to care about us, that is why we personify nature, whether as God, Goddess, Gaia or Mother Nature. But it is an indifferent process as far as our emergent hopes and dreams. Although we as a species, seem to be integral to the universe our place in it has been shaped not by some divine plan but by the process of evolution on the genetic material we carry within us. Whether we as a species survive depends on the fitness of our attributes to the world as we find it minute-by-minute, eon-by-eon.

There is no balance of nature either. There is only the process. It is not balanced because it is always evolving. Energy goes in, is processed into more living things through growth and reproduction. Genetic information is transmitted. Mutations occur. Selections are made. Life goes on or not. The mystery is that there is no mystery.

We romanticize nature. We think of a Turner landscape. We imagine a tropical paradise. It is nonsense. Some religions, such as Buddhism and Jainism have injunctions against killing any living thing this is as unnatural as the Christian concept of having dominion over all creatures. Killing is a part of the process of life. It is a necessary survival tactic. We kill to eat. Even vegetarians kill plants.

By sentimentality about nature I am referring to the belief that we can somehow be harmony with nature. For example, when talk about a reverence for "Mother Nature". This sort of thing is suitable for a greeting card but for the serious business of restoring health to the planet it is totally inadequate.

The World is Illusion

Buddhists and Hindus believe that the world we experience through the senses is a distortion or illusion of the real world which we are not able to experience directly. At one level this does not conflict with the western concept of mind which experiences reality indirectly through the senses; although when I trip in the dark, the world seems real to me. Some physicists hypothesize that the world is a holographic image of reality. The difference however is that the Eastern traditions are geared to transcend the illusion of our senses to reach actual reality.

Whether or not the reality we sense is the real reality is beside the point as far as our treatment of the Biosphere is concerned. It does not matter to us as humans on planet earth what reality is or is not. What matters is that we deal with the empirical facts of the world; Biosphere is real enough and our lives depend on it.

The Magical Thinking of Prayers and Incantations

If you pray for rain and it subsequently rains some might believe that your prayer was answered. But for that to be true there would have to be someone or something that heard your prayer and that entity would have to have the power to override the laws of physics. A more rational and simpler explanation would be that is was a coincidence. Violating the laws of physics would have serious consequences for the universe, in terms of cause and effect. Miracles are, effects without a physical cause. This is contrary to what Einstein proved; the universe is deterministic at the macroscopic scale; physical events must have a physical cause.

Living in Harmony with Nature

The concept of living in harmony with nature is nonsense. There is no spirit of life with which we can commune. Life is a process. It is like trying to have a conversation with a tiger. You might believe you eve all sorts of mystical nonsense about the tiger, but to the tiger you are simply an easy meal. Nature is about survival, not romantic fantasies no matter how enjoyable. Animals survive by taking not asking. In the wild the polite animal is dinner. Evolution only rewards the survivor. The loser has his skeleton exhibited in a museum.

My God is the only God

It is a human characteristic to believe that everyone would be much better off if they had the same beliefs as I do. This is probably a survival adaptation to keep us from second guessing ourselves at every turn. The downside is that theological arguments have caused the most brutal wars in human history. Unlike political wars, religious wars recognize no innocents. Children are slaughtered as quickly as adults.

The American Civil War was in large measure a religious war. The theological argument was whether God condoned slavery or not. The South said yes, and the north said no. Both sides made liberal use of scripture to justify their positions. In the end, the issue was settled by force of arms at the cost of over 650,000 lives.

Humans are innately good: Humans are innately bad

Humans are not inherently good or bad, they are just sentient primates with their own set of evolutionary adaptations. How we use these adaptations is a whole other question. In terms of the biosphere we can maintain it or destroy it, the choice is ours.

The wisdom of Free Markets

For some, free market capitalism is God's law. Imbued with an intelligence beyond that of individuals to properly allocate resources. We believe that somehow everything will be ok if we just let the "market" work it out. That is simply not true when it comes to protecting natural infrastructure. Quite the opposite is true. The market values the pieces of native ecosystems rather than the ecosystems themselves. Lumber, food, fiber, minerals and metals are the commodities from the land or from under it. The fact that the native ecosystems which occupied the land provided oxygen and help regulate the earth's temperature does not figure into the equation.

The restoration of native ecosystems comes down to land use decisions. In this country there is a particularly strong belief in property rights, which are held by some to be sacred. The idea that a piece of ground should be regulated for the purpose of protecting a native ecosystem can anger some to the point of insurrection. The fierce defense of private property is very often tied to the myth of humans having dominion over the world. Nevertheless, every private land use decision has implications beyond the boundaries of a person's property. Decisions about farming, education, transportation, religion, heritage, burials, dwellings, manufacturing, all have impacts on the Biosphere.

When property rights are believed to be granted by God, landowners become empowered to do whatever they like with their land; log it, mine it, pave it, even put it in trucks and cart it off etc. A corollary of this belief is that that governmental regulations regarding land use such as zoning laws, restrictions against filling wetlands, and protection for endangered species, are little more than interference with a God given right.

Those who argue against ecosystem restoration and protection also can make use of other moral imperatives of our society. The idea of economic growth has taken on the trappings if not the actuality of a moral imperative in much of the world. It is particularly resonant in the United States. The basic notion is that our economy must keep growing and any idea of economic sustainability is not practical and may even be unpatriotic.

I am not an economist, but it is obvious to me that the idea of continuous economic growth which relies on every expanding use of energy and resources is by its very nature contrary to the to the sustainability of native ecosystems, because it is the native ecosystems that are the places which must be exploited to maintain economic growth. The constant expansion of drilling for oil and natural gas into more and more native ecosystems is just one example.

When I was working on flood control projects I was forced to try and put ecosystem values into economic terms. That is simply federal policy in the United States. This reflects the deep belief held in this society that economic value trumps the intrinsic value of native ecosystems. But by doing this we ignore the biological value of native ecosystems? For example, the amount of oxygen produced by the forest or its effect on the world's temperature or its sequestration of greenhouse gasses.

ABOUT THE AUTHOR

Who am I and why do I believe I can contribute to understanding and stopping our rush to extinction? To begin with I am both a scientist and religious person. I am a Unitarian Universalist, a liberal religion with no set theology and no requirement to believe in a particular God or in fact any God at all; perfect for someone like me who does not want to be told what to believe. As part of my spiritual journey I have read and studied the world's major religions and I believe that I have an informed opinion on their beliefs regarding the subjects addressed in this book.

I am also a scientist and ecosystem restorationist. I have a bachelor's and master's degrees in Zoology and Wildlife Ecology respectively from the University of Florida and a PhD in Animal Nutrition from the Pennsylvania State University. I have done original research as well as extensive field work. I spent my career as a wildlife biologist for the USDA – Natural Resources Conservation Service.

I worked on many different aspects of natural resources protection including Environmental Impact Studies, wildlife habitat improvement and ecosystem restoration. For the last 10 years or so of my career I was the lead ecologist on an inter-agency effort to restore salt marshes and other native ecosystems in Coastal New Hampshire. My practical experience in ecosystem restoration includes major ecosystem restoration projects that I initiated in New Hampshire. These projects required the cooperation of many government agencies, not for profit environmental organizations, public officials, volunteers and landowners.

I taught ecosystem restoration for my former agency and in many other venues. Based on a nomination by my peers, I received a National Wetlands Award in 2006 for my interagency leadership in the restoration of native ecosystems. I received this award in a ceremony in the Capital Building in Washington DC. The National Wetlands Awards are administered by the Environmental Law Institute, and supported by the US Fish and Wildlife, The Environmental Protection Agency, other federal agencies.

During my career, I developed methods for evaluation of wetland functions and values. These methods were intended to educate local officials about the value of wetlands in within their jurisdiction and to help them make sound land use decisions affecting them. At least one of these methods has been updated and is still in use today. I also contributed to a national method for wetland evaluation as part of an inter-agency effort lead by the US Army Corps of Engineers.

Made in the USA
Middletown, DE
22 November 2021